T0181341

Wissenschaftliche Reihe Fahrzeugtechnik Universität Stuttgart

Series Editors

Michael Bargende, Stuttgart, Deutschland

Hans-Christian Reuss, Stuttgart, Deutschland

Jochen Wiedemann, Stuttgart, Deutschland

Das Institut für Fahrzeugtechnik Stuttgart (IFS) an der Universität Stuttgart erforscht, entwickelt, appliziert und erprobt, in enger Zusammenarbeit mit der Industrie, Elemente bzw. Technologien aus dem Bereich moderner Fahrzeugkonzepte. Das Institut gliedert sich in die drei Bereiche Kraftfahrwesen, Fahrzeugantriebe und Kraftfahrzeug-Mechatronik. Aufgabe dieser Bereiche ist die Ausarbeitung des Themengebietes im Prüfstandsbetrieb, in Theorie und Simulation. Schwerpunkte des Kraftfahrwesens sind hierbei die Aerodynamik, Akustik (NVH), Fahrdynamik und Fahrermodellierung, Leichtbau, Sicherheit, Kraftübertragung sowie Energie und Thermomanagement – auch in Verbindung mit hybriden und batterieelektrischen Fahrzeugkonzepten. Der Bereich Fahrzeugantriebe widmet sich den Themen Brennverfahrensentwicklung einschließlich Regelungs- und Steuerungskonzeptionen bei zugleich minimierten Emissionen, komplexe Abgasnachbehandlung, Aufladesysteme und -strategien, Hybridsysteme und Betriebsstrategien sowie mechanisch-akustischen Fragestellungen. Themen der Kraftfahrzeug-Mechatronik sind die Antriebsstrangregelung/Hybride, Elektromobilität, Bordnetz und Energiemanagement, Funktions- und Softwareentwicklung sowie Test und Diagnose. Die Erfüllung dieser Aufgaben wird prüfstandsseitig neben vielem anderen unterstützt durch 19 Motorenprüfstände, zwei Rollenprüfstände, einen 1:1-Fahrsimulator, einen Antriebsstrangprüfstand, einen Thermowindkanal sowie einen 1:1-Aeroakustikwindkanal. Die wissenschaftliche Reihe „Fahrzeugtechnik Universität Stuttgart" präsentiert über die am Institut entstandenen Promotionen die hervorragenden Arbeitsergebnisse der Forschungstätigkeiten am IFS.

Edited by
Prof. Dr.-Ing. Michael Bargende
Lehrstuhl Fahrzeugantriebe
Institut für Fahrzeugtechnik Stuttgart
Universität Stuttgart
Stuttgart, Deutschland

Prof. Dr.-Ing. Hans-Christian Reuss
Lehrstuhl Kraftfahrzeugmechatronik
Institut für Fahrzeugtechnik Stuttgart
Universität Stuttgart
Stuttgart, Deutschland

Prof. Dr.-Ing. Jochen Wiedemann
Lehrstuhl Kraftfahrwesen
Institut für Fahrzeugtechnik Stuttgart
Universität Stuttgart
Stuttgart, Deutschland

More information about this series at http://www.springer.com/series/13535

Antonino Vacca

Potential of Water Injection for Gasoline Engines by Means of a 3D-CFD Virtual Test Bench

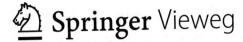

Antonino Vacca
Institute of Automotive Engineering
(IFS), Chair in Automotive Powertrains
University of Stuttgart
Stuttgart, Germany

Zugl.: Dissertation Universität Stuttgart, 2020

D93

ISSN 2567-0042   ISSN 2567-0352   (electronic)
Wissenschaftliche Reihe Fahrzeugtechnik Universität Stuttgart
ISBN 978-3-658-32754-5   ISBN 978-3-658-32755-2   (eBook)
https://doi.org/10.1007/978-3-658-32755-2

This Springer Vieweg imprint is published by the registered company Springer Fachmedien Wiesbaden
GmbH part of Springer Nature.
The registered company address is: Abraham-Lincoln-Str. 46, 65189 Wiesbaden, Germany

,,Work is what you make it!"
Picture of "Felix the cat", sliding with skates made of sponges on a soapy
floor to be cleaned. Drawn by Gino Ciotola and hanging in the room where I
grew up in Isernia (Italy)

Preface

This work was realized during my tenure as a research associate at the Institute of Automotive Engineering (IFS) of the University of Stuttgart under the supervision of Prof. Dr.-Ing. M. Bargende.

My deep gratitude goes to Prof. Dr.-Ing. Michael Bargende for providing the opportunity for this work. His scientific supervision and the exciting conversation about engine thermodynamics were particularly valuable for me.

I also would like to thank Prof. Federico Millo from Politecnico di Torino for the co-referent of my work. From him, I got the principles of internal combustion engines, and he was able to pass me on his passion and some of his most valuable knowledge.

I am extremely grateful to my team leader Dr.-Ing. Marco Chiodi for the challenge he gave me. I want to thank him for the most interesting engine discussions ever, for his kindness and to be an example to me in my work and private life. He helped me identify my strengths and contributed positively to my personal and professional development.

My appreciation also extends to my closest colleagues Francesco, Marcel, Rodolfo, Edoardo, Simon, Hubert, Andreas and Oli. Each one of them has contributed to my professional and personal growth.

Thank you to Mrs. Maike Gern and Mr. Tim Franken for the great cooperation during the project that made this work possible. I would like also to thank Dr.-Ing André-Casal Kulzer of Porsche AG for the technical support as chairman and as valuable thermodynamic guide.

Since the most beautiful part of something always comes to the end, I would like to dedicate this work to my mum Fabia, my dad Paolo, my sister Eleonora, my granddad Antonino, and to my sweet Greta, because without each of them it would make no sense to get a result. Thank you for inspiring me, for making me happy and proud to have such a family.

Antonino Vacca

Stuttgart

Contents

Figures

Tables

Abbreviations

CO_2	Carbon dioxide
T_3	Exhaust gas temperature before turbine
ε	Compression ratio
p_3	Exhaust gas pressure before turbine
$p_{2.2}$	Intake pressure before cylinder
0D	Zero-dimensional
1D	One-dimensional
3D	Three-dimensional
a.FTDC	after firing top dead center
b.FTDC	before firing top dead center
BDC	Bottom dead center
BEV	Battery electric vehicles
BMEP	Break mean effective pressure
CA	Crank angle
CAD	Computer-aided design
CFD	Computational fluid dynamics
CO	Carbon monoxide
CPU	Central processing unit
DI	Direct injection
DISI	Direct injection spark ignition
DNS	Direct numerical simulaion
DOI	Duration of injection
Dp	Particles size class
DWI	Direct water injection
DWI-C	DWI central
DWI-L	DWI lateral

ECU	Electronic control unit
EGR	Exhaust gas recirculation or residual burned gas of previous cylce
EIVC	Early intake valve closure
EIVO	Early intake valve opening
EOI	End of injection
FKFS	Forschungsinstitut für Kraftfahrwesen und Fahrzeugmotoren Stuttgart
FTDC	Firing top dead center
FVV	Forschungsvereinigung Verbrennungskraftmaschinen
HC	Unburned hydrocarbon
HCCI	Homogeneous charge compression ignition
HOV	Heat of vaporization
ICE	Internal combustion engines
IFS	Institut für Fahrzeugtechnik Stuttgart
IMEP	Indicated mean effective pressure
IP	Ignition point
ISFC	Indicated specific fuel consumption
IVC	Intake valve closure
IVK	IVK targeting
IVK	Institut für Verbrennungsmotoren und Kraftfahrwesen
IVK T	IVK tumble targeting
IWI	Indirect water injection
IWI-IM	IWI intake manifold
IWI-VS	IWI valve selective
KRAT	Knocking ratio
LES	Large eddy simulation
LIF	Laser induced fluorescence
M1	Miller valve profile 1
M2	Miller valve profile 2

MFB10-90	Combustion duration corresponding to 10-90% burned mass fraction
MFB50	50% Burned mass fraction
N	Particles number concentration
NEDC	New european driving cycle
NOx	Oxides of nitrogen
OP1	Operating point 1
OP2	Operating point 2
OP3	Operating point 3
PDA	Phase doppler analyzer
PFI	Port fuel injection
PMAX	Maximum pressure
PNC	Particulate number concentration
PRF	Primary reference fuel
RANS	Reynolds average navier stokes
RDE	Real driving emissions
RH	Initial relative humidity of the air
RON	Research octane number
rpm	Revolutions per minute
RTS95	Standardized random test and aggressive driving style 95
SI	Spark-ignition
SMD	Sauter mean diameter
SOI	Start of injection
SOIW	Start of water injection
Std.	Standard
Std.100	Standard 100 bar
TDC	Top dead center
TKE	Turbulent kinetics energy
TRF	Ternary reference fuel
TU	Technischen Universität

TWC	Three way catalyst
WLTC	Worldwide harmonized light duty test cycle
WOT	Wide open throttle
WPC	Real working process calculation

Symbols

Latin Letters

A	Area	m^2
K	Wrinkling factor	-
S_L	Laminar flame speed	m/s
S_T	Turbulent flame speed	m/s
c_f	Long-range proces term in conservation equation	-
c_p	Specific heat at constant pressure	J/kgK
F	Extensive variable in conservation equation	-
f	Density / intensive variable in conservation equation	-
G	Gravitational energy	J
H_E	Exhaust enthalpy	J
h_f	Specific heat of formation	J/kg
H_I	Intake enthalpy	J
H_L	Leakage enthalpy	J
U	Internal energy	J
h_{tc}	Specific thermo-chemical enthalpy	J/kg
h	Specific enthalpy	J/kg
p	pressure	Pa
u	Specific internal energy	J/kg
v	Specific volume	m^3/kg
i	Stroke factor	-
$\vec{j}$	Diffusiv mass flux	$kg/m^2 s$
k	Mass specific turbulent kinetic energy	$m^2/s^2/kg$
l	Characteristic length	m
M	Molar mass	kg/kmol
m	Mass	kg
m_B	Fuel mass	kg
m_C	Cylinder mass	kg

$\dot{m}$	Mass flow	kg/s
n	Engine speed	rpm
Oh	Ohnesorge number	-
P_e	Effective power	W
pme	Mean effective pressure	Pa
$\bar{\bar{P}}$	Second order stress tensor	N/m^2
$p\bar{\bar{I}}$	Momentum due to pressure	N/m^2
Q_B	Fuel heat release	J
Q_W	Wall heat transfer	J
R_s	Specific gas constant	J/kgK
Re	Reynolds number	-
s_f	Source or sink term in conservation equation	-
T	Temperature	K
t	Time	s
$\vec{V}$	Diffusiv fluid velocity	m/s
V	Volume	m^3
$\vec{v}$	Fluid velocity	m/s
V_{cyl}	Engine displacement	m^3
W	Work	J
w	Mass fraction	kg/kg
We	Weber number	-

Greek Letters

α	Spray jet angle	deg
ε	Geometric compressure ratio	-
η	Dynamic viscosity	Ns/m^2
η_f	Droplet dynamic viscosity	Ns/m^2
γ	Hollowcone spray angle	deg
α_{st}	Air to fuel ratio stoichiometric	-
λ	Air to fuel ratio	-
Ω	Fixed volume	m^3
$\Delta\Omega$	Surface of volume Ω	m^2
ω_i	Molar formation rate of species i	kmol/m^{3}s
φ	Crank angle	deg
φ	Spray azimuth angle	deg

$\vec{\Phi}_f$	Flux of the general density variable f in the conservation equations	-
$\overline{\overline{\Pi}}$	Momentum due to viscous forces	N/m^2
ρ	Density	kg/m^3
ρ_g	Density of the droplet surrounding gas	kg/m^3
ρ_i	Density of species i	kg/m^3
σ	Surface tension	N/m
σ_f	Droplet surface tension	N/m
Δt	Time step	s
τ_{ID}	Ignition delay time	s
τ	Characteristic droplet breakup time	s
θ	Spray zenith angle	deg

Abstract

CO_2 emission constraints lead to further investigations of technologies to lower knock sensitivity of gasoline engines, the main limiting factor to increase engine efficiency and thus reduce fuel consumption. Moreover, the RDE cycle demands higher power operation, where component protection requires fuel enrichment. The engine must operate at stoichiometric conditions to achieve high efficiency with lower fuel consumption, and to be able to convert emission efficiently.

The reasons for water injection
Among others, water injection is one of the most promising technologies to improve engine combustion efficiency, by reducing knock occurrences, controlling exhaust gas temperature and allowing the engine running at stoichiometric conditions over the whole engine map. The implementation of water injection can be realized through different engine layouts and injection strategies. Besides, the results of injecting water are strictly dependent on the base engine features and load point. Therefore, engine virtual development becomes essential to study many engine configurations.

In this dissertation, indirect (IWI) and direct water injection (DWI) strategies allow to increase the engine knock limit and to reduce exhaust gas temperature, at three different operating points.

QuickSim advantages
The investigation was conducted with the 3D-CFD-Tool QuickSim. The implementation of water physics into QuickSim represented the first part of this work. Since the many possible water injection engine configurations to be tested, the reproduction of the water effects into the engine called for a fast 3D-CFD simulation tool that would have been able to produce time-efficient computations. For such reason, the description of the working fluid is based on six scalars, *air*, *fuel*, *EGR* and the respectively" burned "scalars. The addition of a new external medium (water) brought to program three new entities, *water*, *EGR Water* (water coming from previous cycles) and *Burned Water* (water taking part to the combustion process, vaporized and crossed over by

the flame front). Accordingly to this methodology, the scalars are not directly related to specific chemical species, but they are optimized to identify and describe the working fluid in its main groups of interest during all the phases of the engine operating cycle. By coupling this working fluid definition, and accurate numerical description of real fuels, a new 3D-knock-model was also implemented, which was validated experimentally also with water injection measurements.

Experiments and simulation

Measurements were carried out on a single-cylinder engine, with indirect and direct lateral water injections, which were used to validate the simulations. The introduction of water directly in the cylinder was operated with a conventional direct gasoline injector, with asymmetrical spray pattern. The results obtained at the test bench using this gasoline device with water suggested the need for the development of a water-optimized spray targeting. For this purpose, a new injector targeting was studied through the support of 3D-CFD simulations, after calibration of the simulation through the base experimental data. The water-developed injector targeting, together with the optimization of the water injection strategy, was led through the virtual screening of different configurations and mainly by combining several starts of water injection, injector positions, injection pressures and water to fuel ratios.

Virtual test bench for the investigation of Miller strategies and DWI

The analysis is marked out as *3D-CFD Virtual Test Bench* because of the massive amount of configurations tested through simulations and for the similarity to the experimental methodology usually adopted for the investigation of the engine thermodynamics at the test bench. The set-up of the Virtual Test Bench was then exploited to evaluate earlier spark advance for knock onset, together with higher compression ratio (ε) in combination with Miller valve strategy and DWI. The results showed the essential contribution of fast 3D-CFD simulations for increasing engine efficiency through water injection and proposing a new approach for the future power-train engine development process.

Spray model

In parallel with the implementation of water thermodynamics in QuickSim, it was necessary to develop a procedure that would allow the coupling of optical measurements conducted on injector sprays with CFD simulations. Within the *FVV-Project n.1256*, a measurement campaign was carried out at a spray test bench, to characterize the difference in the spray formation, when injecting water instead of gasoline, at low (5-10bar) and high (50-200 bar) injection pressure. A new methodology for coupling *Mie-scattering/shadow imaging* and 3D-CFD simulations was developed with the intent of increasing the reliability of the injection simulations and with the additional objective of investigating the spray characteristics, together with the virtual engine development. This approach led to enhance the settings of the breakup model and particularly the calibration of the droplets initialization domain, which in turn means the calibration of the spray pattern and penetration within the combustion chamber. The developed procedure represents a global approach for the spray geometry analysis and therefore for the design of new sprays, accomplished by taking into account together mixture formation (in a real engine flow field), influence on the combustion process and knock onset but as well wall impingement issues.

Set-up of the virtual test bench
The numerical implementation of water injection in the 3D-CFD tool QuickSim and the spray investigation were the fundamentals of the further full engine analysis. In a second step, the mesh of the single-cylinder engine at the test bench was built considering the whole system, from the auxiliary pressure vessel, up to the exhaust muffler. The advantage was the stability of the system, the possibility of setting environmental boundary conditions and the prospect to investigate as well IWI strategies, by being aware of the flow field of the intake/exhaust system and eventually of scavenging processes.

Limiting factor for water evaporation
The first problem was the lower cooling efficiency of injecting water to theoretical assessment. Exploiting the thermodynamic effect of water injection was not completely possible because of the engine operating parameters that were limiting water evaporation or imposing injection strategy constraints to avoid wall wetting or oil dilution. Besides, the very high water evaporation enthalpy was responsible for huge local temperature reduction, preventing, as a result, the evaporation of further water droplets, due to the decrease of the pressure of saturation. For these reasons, the injection strategies were re-oriented in order to enhance as much as possible the air entrainment into the spray and thus the water evaporation.

Operating point 1

Meaningful was the adoption of DWI for *Operating point 1* (2000 rpm 20 bar IMEP) with respect to IWI. This load point was characterized by a much-delayed ignition point (1°CA b.FTDC), because of the high knock sensitivity with base calibration. Especially IWI was not able to provide a sufficient charge cooling effect before ignition point, and therefore, its influence in reducing knock tendency was lower. Independently of the applied injection strategy, by keeping 30% water to fuel ratio, IWI was not effective for improving 50% Burned mass fraction (MFB50) at this load point. On the other side, almost all DWI tested strategies led to advance the combustion phasing with two separate mechanisms. First, the charge cooling effect raised by injecting water directly into the cylinder, with a favourable flow field for water evaporation, together with smaller droplets and higher jet velocities. Secondly, the direct water injection could be used to improve the charge motion within the cylinder. In this sense, developing a specific spray angle and spray pattern the engine *Tumble* could be fostered and therefore, the turbulent kinetic energy boosted. This second effect improved mixture formation and triggered faster combustion, making the flame front running over the end-gas region sooner and finally enhancing knock resistance. DWI could be even employed in some case for increasing in-cylinder pressure, without advancing the spark timing, leading to higher indicated efficiency but respecting mechanical constraints. In detail, at *Operating point 1*, the combination of DWI at 200 bar with water-developed spray targeting, it allowed to cool down the charge, to improve mixture formation, due to the higher turbulence, and the spark advance could be moved earlier, due to the enhanced knock resistance. MFB50 moved from 21 to 12 °CA a.FTDC, shortening combustion duration and leading to increase the engine indicated efficiency up to 12%. The load point was always reached with the same injected fuel, keeping stoichiometric mixture, which means that more efficient combustion resulting in higher torque.

Operating point 1 with Miller strategy

Finally, the experimentally validated Virtual Test Bench was used to study the possible rise of ε, in combination with Miller strategy and DWI for *Operating point 1*. The Miller lift was generated keeping the same valve acceleration and considering different maximum lifts (from 25 up to 50% max lift reduction to the base cam profile). The reduced intake air mass, combined with the water cooling down effect led to adopt higher ε (12 : 1 instead of 10.75 : 1), without overcoming the knock threshold, adding to the already enhanced indicated efficiency, a further improvement margin due to the longer expansion stroke.

Operating point 2
Considering *Operating point 2* (2500 rpm 15 bar IMEP), the injection of water, both indirectly or in the combustion chamber, produced no advantages in terms of combustion phasing and therefore no fuel consumption reduction. At *Operating point 2*, the base engine calibration already had an earlier spark advance and no improvement margins were found. On the other side the emission benefits were clear, considering NOx formation. For the soot production instead, it was found that the start of water injection played a key role.

Operating point 3
About IWI strategies, the most important result was achieved for reducing exhaust gas temperature before the turbine. For rated power point of the engine (4500 rpm 20 bar IMEP, *Operating point 3*) IWI showed higher potential for lowering T_3, in comparison to DWI. This result was obtained by moving the injector from the intake manifold directly on the intake port. No clear influence among 5 and 15 bar injection pressure was highlighted. Late injection timing was able to compensate for lower injection pressure in terms of T_3 reduction. Overall, both IWI and DWI allowed avoiding fuel enrichment completely with 30 % water to fuel ratio. The rated power operating point could be optimized, decreasing fuel consumption by up to 12 % to the base calibration. T_3 was reduced at the same time by 85 K if stoichiometric combustion and no water injection was considered, meaning 40 K lower T_3 to fuel enrichment case.

Conclusion
3D-CFD simulations are a powerful and essential tool for an in-depth analysis of water injection effects in gasoline engines. Among these, the Virtual Test Bench, thanks to its ability to cope with new operating parameters and hardware solutions, gives insights and results even more significant for the implementation of the water injection in series engines. The water effects are strongly dependent on engine configuration and particularly on load point.

IWI strategies do not represent a second choice comparing with DWI, for T_3 reduction. By optimizing the water injection strategy, IWI can realize the same T_3 decrease as DWI and even improve it, thanks to the exploitation of the post evaporation process. IWI configurations show potentially less wall impingement and thus a lower oil dilution effect. As a drawback, IWI increases the possibility of intake manifold condensation and uneven water distribution over the cylinders. On the other side, DWI strategies enable to control better in-cylinder temperature enhancing the knock resistance and improving the engine indicated efficiency as for *Operating point 1*. The design of a direct water-optimized injector plays a crucial role in the engine response to water injection. Wider spray targeting, allowed to increase the air entrainment, thus promoting water evaporation, but wall impingement analysis must be simultaneously run.

Kurzfassung

Die Beschränkungen der CO_2-Emissionen führen zu weiteren Untersuchungen von Technologien zur Verringerung der Klopfempfindlichkeit von Ottomotoren, dem wichtigsten limitierenden Faktor zur Erhöhung des Wirkungsgrads und einer damit einhergehenden Senkung des Kraftstoffverbrauchs. Darüber hinaus erfordert der RDE-Zyklus einen Betrieb mit höherer Leistung, bei dem bisher eine Kraftstoffanreicherung zum Schutz der Komponenten erforderlich ist. Um jedoch einen hohen Wirkungsgrad bei geringen Emissionen zu erreichen, sollte der Motor in allen Kennfeldbereichen unter stöchiometrischen Bedingungen betrieben werden.

Die Gründe für die Wassereinspritzung

Die Wassereinspritzung ist eine der vielversprechendsten Technologien zur Verbesserung der Verbrennungseffizienz von Motoren, indem sie das Klopfen reduziert, die Abgastemperatur senkt und den Motor über das gesamte Motorkennfeld unter stöchiometrischen Bedingungen laufen lässt. Die Umsetzung der Wassereinspritzung kann durch verschiedene Motorauslegungen und Einspritzstrategien realisiert werden. Zudem sind die Ergebnisse der Wassereinspritzung streng von den grundlegenden Motoreigenschaften und den betrachteten Lastpunkten abhängig. Daher wird die virtuelle Motorenentwicklung für die Untersuchung vieler Motorkonfigurationen im Hinblick auf die Wassereinspritzung künftig eine bedeutende Rolle spielen.

In dieser Dissertation wurden indirekte (IWI) und direkte Wassereinspritzungsstrategien (DWI) zur Erhöhung der Klopfgrenze des Motors und zur Reduzierung der Abgastemperatur in drei verschiedenen Betriebspunkten betrachtet.

QuickSim Vorteile

Die Untersuchung wurde mit dem 3D-CFD-Tool QuickSim durchgeführt. Die Umsetzung der physikalischen Eigenschaften des Wassers in QuickSim stellt den ersten Teil dieser Arbeit dar. Da die vielen möglichen Konfigurationen von Wassereinspritzungs an Motore getestet werden sollten, erforderte die Reproduktion dieser Stoffeigenschaften im motorischen Kontext erfordert ein schnelles 3D-CFD-Simulationswerkzeug, das zeiteffiziente Berechnungen ermöglicht. Aus diesem Grund basiert die Beschreibung des Arbeitsmediums auf sechs Skalaren: *Luft, Kraftstoff, AGR* und den jeweils " verbrannten" Skalaren. Die Implementierung eines neuen externen Mediums (Wasser) brachte drei neue Entitäten zum Programm, *Wasser, EGR Wasser* (Wasser, das aus früheren Zyklen stammt) und *Verbranntes Wasser* (Wasser, das am Verbrennungsprozess teilnimmt, verdampft und von der Flammenfront überquert wird). Entsprechend dieser Methodik sind die Skalare nicht direkt mit bestimmten chemischen Spezies verbunden, werden aber mit dem Ziel optimiert, das Arbeitsmedium in seinen wichtigsten Komponenten während aller Phasen des Motorbetriebszyklus zu identifizieren und zu beschreiben. Durch die Kopplung dieser Arbeitsfluid-Definition mit der Reproduktion realer Kraftstoffe wurde auch ein neues 3D-Klopfmodell implementiert, das experimentell mit Wassereinspritzmessungen validiert wurde.

Experimente und Simulation

Experimente wurden an einem Einzylinderaggregat mit indirekter und direkter seitlicher Wassereinspritzung durchgeführt, die zur Validierung der Simulationen verwendet wurden. Da es sich bei der direkten Wassereinspritzung um ein konventionelles Benzinventil mit asymmetrischem Strahlbild handelte, legten die am Prüfstand erzielten Ergebnisse die Notwendigkeit der Entwicklung eines wasseroptimierten Strahlbildes nahe. Zu diesem Zweck wurde nach der Kalibrierung der Simulation anhand der experimentellen Basisdaten mit Hilfe von 3D-CFD-Simulationen ein neues Spraybildung untersucht. Das mit Wasser entwickelte Injektor-Targeting wurde zusammen mit der Optimierung der Wassereinspritzstrategie durch das virtuelle Screening verschiedener Konfigurationen und hauptsächlich durch die Kombination verschiedener Wasserinjektionsstarts, Injektorpositionen, Einspritzdrücke und Wasser-Kraftstoff Verhältnisse geführt.

Virtueller Prüfstand zur Untersuchung von Millerstrategien und DWI

Die Analyse wird als *3D-CFD Virtueller Prüfstand* bezeichnet, wegen der großen Anzahl der durch Simulationen getesteten Konfigurationen und aufgrund der Ähnlichkeit mit der experimentellen Methodik, die normalerweise für die Untersuchung der Thermodynamik des Motors auf dem Prüfstand verwendet wird. Der Aufbau des virtuellen Prüfstands wurde dann verwendet, um frühere Zündzeitpunkte in Bezug auf die Klopfneigung zusammen mit einem höheren Verdichtungsverhältnis (ε) in Kombination mit der Miller Ventilstrategie und DWI zu bewerten. Die Ergebnisse zeigten einen wesentlichen Beitrag von schnellen 3D-CFD-Simulationen zur Steigerung der Motoreffizienz durch Wassereinspritzung und einen neuen Ansatz für den zukünftigen Entwicklungsprozess von Ottomotoren.

Spraymodell

Parallel zur Umsetzung der Wasser-Thermodynamik in dem 3D-CFD Tool QuickSim war es notwendig, ein Verfahren zu entwickeln, das die Kopplung von optischen Messungen an Injektorsprays mit CFD-Simulationen ermöglicht. Im Rahmen des *FVV-Projekt n.1256* wurde eine Messkampagne an einem Strahlprüfstand durchgeführt, mit dem Ziel, den Unterschied in der Spraybildung beim Einspritzung von Wasser statt Benzin bei niedrigem (5-10bar) und hohem (50-200 bar) Einspritzdruck zu charakterisieren. Eine neue Methodik zur Kopplung von *Mie-Streuung/Schattenbild* und 3D-CFD Simulationen wurde mit der Absicht entwickelt, die Zuverlässigkeit der Einspritzungsimulationen zu erhöhen und mit dem zusätzlichen Ziel, die Spraycharakteristik zusammen mit der virtuellen Motorenentwicklung zu untersuchen. Dieser Ansatz führte zu einer Verbesserung der Einstellungen des Aufbruchsmodells und insbesondere der Kalibrierung der Tröpfcheninitialisierungsdomäne, was wiederum die Kalibrierung des Spraymusters und der Penetration innerhalb des Brennraums bedeutet. Das entwickelte Verfahren stellt einen globalen Ansatz für die Analyse der Spraygeometrie und damit für die Auslegung neuer Sprays dar. Die gemeinsame Berücksichtigung der Gemischbildung (in einem realen Motorströmungsfeld), deren Einfluss auf den Verbrennungsprozess und auf das Motorklopfen, aber auch Fragen wie Wandbenetzung bewirken einen Fortschritt auf die Qualität der Simulationsergebnisse im Motorentwicklungsprozess.

Aufbau des virtuellen Prüfstands

Die numerische Umsetzung der Wassereinspritzung in QuickSim und die Einspritzuntersuchungen waren die Grundlagen für die weitere vollständige Motoranalyse. In einem zweiten Schritt wurde das Netz des Einzylindermotors am Prüfstand unter Berücksichtigung des gesamten Systems, vom Druckbehälter bis zum Abgasschalldämpfer, aufgebaut. Der Vorteil liegt hier in der Stabilität des Systems, der Möglichkeit, Umgebungsrandbedingungen einzustellen und der Aussicht, auch IWI-Strategien zu untersuchen, indem man das

Strömungsfeld des Ansaug-/Abgassystems und schließlich die Spülvorgänge kennt.

Grenzen der Wasserverdampfung

Eine erste Problematik ist die geringere Kühleffizienz der Wassereinspritzung im Hinblick auf die theoretische Bewertung. Der thermodynamische Effekt der Wassereinspritzung wurde anfangs nicht voll ausgenutzt, da die Betriebsparameter des Motors die Wasserverdampfung einschränkten oder die Einspritzstrategie Beschränkungen auferlegten, um eine Wandbenetzung oder Ölverdünnung zu vermeiden. Darüber hinaus war die sehr hohe Verdampfungsenthalpie des Wassers für eine enorme lokale Temperaturreduktion verantwortlich, wodurch die Verdampfung weiterer Wassertröpfchen aufgrund des abnehmenden Sättigungsdrucks verhindert wurde. Aus diesen Gründen wurden die Einspritzstrategien neu ausgerichtet, um den Lufteintrag in den Spray und damit die Wasserverdampfung so weit wie möglich zu verbessern.

Betriebspunkt 1

Eindeutig sinnvoll war die Untersuchung einer direkten Wassereinspritzung (DWI) für *Betriebspunkt 1* (2000 U/min 20 bar IMEP) im Vergleich zur indirekten Einspritzung(IWI). Dieser Lastpunkt war durch einen stark verzögerten Zündzeitpunkt (1°CA b.FTDC) gekennzeichnet, aufgrund der hohen Klopfempfindlichkeit bei der Basiskalibrierung. Insbesondere die IWI konnte vor dem Zündzeitpunkt keine ausreichende Ladungskühlung bewirken, so dass ihr Einfluss auf die Verringerung der Klopfneigung gering war. Unabhängig von der angewandten Einspritzstrategie war IWI durch die Beibehaltung eines Wasser-Kraftstoff-Verhältnisses von 30% nicht effektiv in der Lage, die Verbrennungsschwerpunktlagen in diesem Lastpunkt zu verbessern. Auf der anderen Seite führten fast alle getesteten DWI-Strategien dazu, die Schwerpunktlage mit zwei getrennten Mechanismen in Richtung früh zu verschieben. Zum Einen wurde der Ladungskühlungseffekt durch direktes Einspritzen von Wasser in den Zylinder erhöht, mit einem für die Wasserverdampfung günstigen Strömungsfeld, zusammen mit kleineren Tropfen und höheren Strahlgeschwindigkeiten. Zweitens konnte die direkte Wassereinspritzung zur Verbesserung der Ladungsbewegung innerhalb des Zylinders genutzt werden. In diesem Sinne könnte die Entwicklung eines spezifischen Strahlwinkels resp. Spraytargetings, das den Motor *Tumble* fördert, hilfreich sein. Damit würde die turbulente kinetische Energie verstärkt werden. Dieser zweite Effekt verbesserte die Gemischbildung und löste eine schnellere Verbrennung aus, wodurch die Flammenfront früher über den Endgasbereich läuft und schließlich die Klopffestigkeit erhöht wird. DWI konnte in einigen Fällen sogar zur Erhöhung des Zylinderdrucks eingesetzt werden, ohne den Zündzeitpunkt zu verändern (und folglich zu einem höheren angegebenen Wirkungsgrad, aber unter Beachtung der mechanischen Zwänge). Im Einzelnen konnte bei *Betriebspunkt 1* durch die Kombination von DWI bei 200 bar mit einem wasseroptimierten Spraytargeting die Ladung abgekühlt werden, um die Gemischbildung aufgrund der höheren Turbulenz verbessert sowie der Zündzeitpunkt aufgrund der erhöhten Klopffestigkeit nach frühverschoben werden. Die Verbrennungsschwerpunktlage wurde von 21 auf 12 °KW n.ZOT verschoben was dazu führte, dass der Motor einen Wirkungsgradgewinn von bis zu 12% anzeigte und sich nebenbei eine kürzere Brenndauer einstellte. Der Lastpunkt wurde immer mit der gleichen eingespritzten Kraftstoffmasse erreicht, wobei das stöchiometrische Gemisch beibehalten wurde Folglich führt die effizientere Verbrennung zu einem höheren Drehmoment.

Betriebspunkt 1 und Millerstrategie

Schließlich wurde der experimentell validierte virtuelle Prüfstand zur Untersuchung des möglichen Anstiegs der ε in Kombination mit der Miller-Strategie und dem DWI für *Betriebspunkt 1* verwendet. Die Miller-Hubkurve wurde unter Beibehaltung der gleichen Ventilbeschleunigung und unter Berücksichtigung verschiedener maximaler Hübe (von 25 bis zu 50% maximaler Hubreduzierung in Bezug auf das Basisnockenprofil) erzeugt. Die reduzierte Einlassluftmasse, kombiniert mit dem Wasserkühlungseffekt, führte zu höheren Verdichtungsverhältnis (12 : 1 statt 10.75 : 1), ohne die Klopfschwelle zu überschreiten, was zu der bereits verbesserten Effizienz eine weitere Verbesserung aufgrund des längeren Expansionshubs bewirkte.

Betriebspunkt 2

Unter Berücksichtigung von *Betriebspunkt 2* (2500 U/min 15 bar) brachte die Wassereinspritzung, sowohl indirekt als auch direkt in den Brennraum, keine Vorteile hinsichtlich der Verbrennungsschwerpunktlage und somit keine Reduzierung des Kraftstoffverbrauchs. Bei *Betriebspunkt 2* war der Basismotor bereits mit einem früheren Zündzeitpunkt kalibriert, und es wurden keine Verbesserungsmöglichkeiten gefunden. Auf der anderen Seite waren die Emissionsvorteile unter Berücksichtigung einer verringerten NOx-Bildung klar. Für die Rußproduktion wurde stattdessen festgestellt, dass der Beginn der Wassereinspritzung eine Schlüsselrolle spielte, und genau auf das jeweilige Brennverfahren abgestimmt werden muss.

Betriebspunkt 3

In Bezug auf die IWI-Strategien wurde das wichtigste Ergebnis zur Reduzierung der Abgastemperatur bevor Turbine. Für den Nennleistungspunkt des Motors (4500 U/min 20 bar IMEP, *Betriebspunkt 3*) zeigte die Ansaugrohr Wassereinspritzung ein höheres Potenzial zur Senkung von T_3, im Vergleich zur Direkt Wassereinspritzung. Dieses Ergebnis wurde erreicht, indem die Einspritzdüse vom Ansaugrohr direkt auf das Einlassventil bewegt wurde. Es wurde kein deutlicher Einfluss zwischen 5 und 15 bar Einspritzdruck deutlich. Der späte Einspritzzeitpunkt konnte den niedrigeren Einspritzdruck im Sinne einer T_3-Reduktion ausgleichen. Insgesamt konnten sowohl indirekt als auch direkt Wassereinspritzungsstrategien eine ansonsten nötige Kraftstoffanreicherung mit einem Wasser-Kraftstoff-Verhältnis von 30% an Nennleistungspunkt vermeiden. Der Betriebspunkt der Nennleistung konnte optimiert werden, wodurch der Kraftstoffverbrauch um bis zu 12% gegenüber der Basiskalibrierung gesenkt werden konnte. T_3 wurde gleichzeitig um 85 K reduziert, wenn die stöchiometrische Verbrennung und keine Wassereinspritzung berücksichtigt wurde. Somit ist die Abgastemperatur um 40 K geringer als in Bezug auf den Betrieb im gleichen Punkt mit Anfettung.

Schlussfolgerung

3D-CFD-Simulationen sind ein leistungsstarkes und wesentliches Werkzeug für eine detaillierte Analyse der Wassereinspritzeffekte in Ottomotoren. Unter diesen gibt der virtuelle Prüfstand dank seiner Fähigkeit, mit neuen Betriebsparametern und Hardwarelösungen umzugehen, Erkenntnisse und Ergebnisse, die für die Umsetzung der Wassereinspritzung in Serienmotoren wichtige Erkenntnisse mit sich bringen. Die auftretenden Effekte beim Einsatz von Wassereinspritzung sind stark von der Motorkonfiguration und insbesondere vom Lastpunkt abhängig.

IWI-Strategien dürfen im Vergleich zum DWI nicht als zweite Wahl zur Abgastemperatur -Reduzierung betrachtet werden. Durch die Optimierung der Wassereinspritzstrategie kann IWI die gleiche T_3-Reduktion wie DWI realisieren und dank der Ausnutzung des Nachverdampfungsprozesses sogar verbessern. IWI Konfigurationen zeigen potenziell weniger Wandbenetzung und damit einen geringeren Ölverdünnungseffekt. Als Nachteil erhöht die IWI die Möglichkeit der Kondensation im Ansaugrohr und der ungleichmäßigen Wasserverteilung über die Zylinder. Auf der anderen Seite ermöglichen DWI-Strategien eine bessere Steuerung der Zylindertemperatur, wodurch der Klopf-schutz und der Wirkungsgrades des Motors verbessert werden, wie es für *Betriebspunkt 1* realisiert wurde. Die Konstruktion eines direkten wasseroptimier-ten Injektors spielt eindeutig eine Schlüsselrolle für das zeitliche Ansprechver-halten des Motors auf die Wassereinspritzung. Eine breitere Ausrichtung des Sprühstrahls erlaubt es, den Lufteintrag zu erhöhen und damit die Wasserver-dampfung zu fördern. Gleichzeitig muss wiederum eine Analyse der Wandbe-netzung durchgeführt werden.

1 Introduction

Today water injection is investigated for turbocharged gasoline engines to improve engine efficiency over the whole operating range [28]. Several studies on the effect of ethanol-fuel and water demonstrated how to lower down the intake temperature and to decrease the Knock sensitivity. The reason is essentially their higher heat of vaporization (HOV) and particularly water with respect to commercial gasoline [15]. Indeed, one result of water injection is the reduction of in-cylinder gas temperature, allowing for earlier centre of combustion and a lower exhaust gas temperature before turbine (T_3), avoiding fuel enrichment for component protection. Therefore, water adoption endorses engine operations at higher indicated mean effective pressure (IMEP) with a low Knock probability.

Hermann et al. found for indirect water injection (IWI) at 5000 rpm that the 50 % burned mass fraction (MFB50) can be advanced by 5 °CA, the fuel enrichment reduced of 16.5 %, which is a fuel-saving of up to 18 % to run the same load point [24]. For direct water injection (DWI) in the cylinder, similar results are expected but with the advantage of water consumption reduction [25]. Further, Hoppe et al. showed 8 % engine efficiency improvement with water injection by increasing the geometrical compression ratio (ε) [26]. The use of increased compression ratio and modified engine map calibration reveals the potential to reduce fuel consumption, keeping the same performance. For WLTC cycles, the possible fuel savings is 4 % moving to a higher value (7 %) during more demanding cycles such as RTS 95 [26]. De Bellis et al. indicated a complete reduction of knocking combustion for 17 % water to fuel-ratio (W/F) using IWI strategies [14]. They stated a reduction of brake specific fuel consumption (BSFC) of up to 15-20 % for the investigated turbocharged engine. Franzke et al. investigated IWI and DWI for a four-cylinder gasoline engine operated at RDE conditions [17]. The efficiency of the engine could be improved by increasing the compression ratio and reducing the fuel enrichment at high power conditions. The DWI has shown to perform better for improving efficiency. The group of Heinrich et al. presented the benefits of DWI regarding fuel consumption and showed an improvement of 2-3 % to indirect strategies [23].

Limiting factors for applying water injection is the higher system complexity, the needed space to install the system and the necessity to refill water additionally to fuel. Hence, technologies are investigated that recover water from different engine peripheries and reuse it for IWI and DWI. Franzke et al. showed that under RDE operating conditions, enough water is recovered for water injection [17]. Sun et al. investigated different water recovery systems [47]. They stated that water recuperation is most beneficial after exhaust gas recirculation cooler (EGR) or after charge air cooler with recovery efficiencies between 50 % and 94 %.

The thermodynamic effects of water injection are not fully exploited because of the engine operating parameters that limit water evaporation or impose injection strategy constraints to avoid wall wetting or oil dilution. This work aims to extend the knowledge of water thermodynamic effects with an in-depth analysis of IWI and DWI using a novel virtual 3D computational fluid dynamics (CFD) test bench. The study of water effects was realized by implementing water physics within the 3D-CFD-tool QuickSim [11] and by validating the models through the experiments carried out at a single-cylinder engine test bench installed at Technischen Universität (TU) Berlin (Institut für Fahrzeugantriebe).

1.1 The performance of water injection to fulfil RDE cycles

Future efficiency improvement of direct injection spark ignition (DISI) engines, as well as gaseous and particulate emission reduction require innovative approaches. With the new *Real driving emissions (RDE)* legislation for light-duty vehicles, the focus shifts towards dynamic test cycles with respect to other driving profiles such as *New european driving cycle (NEDC)* and the currently adopted one the *Worldwide harmonized light duty test cycle (WLTC)*. Figure 1.1 presents the operating points that a medium weight class vehicle (1500 kg) has to cover when driving over different cycle profiles and particularly NEDC, WLTC and RDE [18] [19].

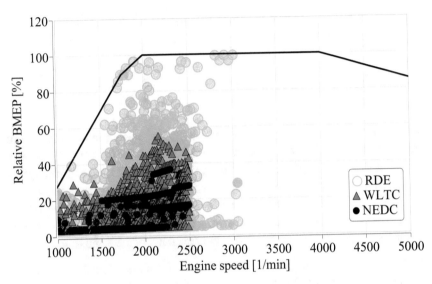

Figure 1.1: Engine map for medium weight class vehicles with operating points for NEDC, WLTC and RDE driving cycle.

It is useful to classify the driven load points into power bins or torque class like reported in table 1.1. The cycles mentioned above could be divided into three relative break mean effective pressure (BMEP) ranges with respect to the considered engine maximum torque. It can be remarked that no load points were covered during NEDC, when analysing the third interval, which represents the load points from 50 to 100 % relative BMEP. For WLTC just 4 % of the load points are comprised in the third torque interval, while up to 16 % of the load points are in this interval when the vehicle drives on RDE. By comparing the share of the third torque class WLTC and RDE the high load points growth of 12 % in case of real driving cycle.

Table 1.1: Classification of the load intervals between NEDC, WLTC and RDE

Torque Interval [#]	Relativ BMEP Interval [%]	NEDC [%]	WLTC [%]	RDE [%]
1	0 - 10	77	59	54
2	10 - 40	23	37	30
3	40 - 100	0	4	16

It is clear that, when the new legislation will take place, the optimization of the high-load part of the engine map becomes essential for controlling emission and fuel consumption. Also, the increase of BMEP for gasoline engines demands for higher Knock resistance and exhaust gas temperatures control, especially considering boosted engines. The use of water injection can reduce both Knock probability and exhaust gas temperature [28] enabling exhaust components protection and efficiency improvement at the same time.

Moreover, as it will happen, in the next Euro 7 emission legislation, exhaust gas products such as CO will be strictly regulated as today is not the case. The onset of such limit would completely forbid the practice of fuel enrichment for component protection. For such reasons, there is a growing interest in the implementation of water injection technology to make still DISI engines attractive.

1.2 Thermodynamic of water injection

1.2.1 Water injection concept for SI-engines

The introduction of water in the engine results in the modification of multiple and interdependent operating parameters [24] [40]. An extension of these bonds is sketched in figure 1.2.

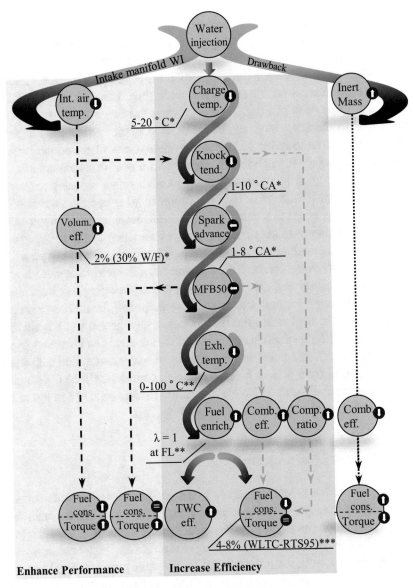

Figure 1.2: Effects of water injection on engine operating parameters and particularly for efficiency enhancement (left) or performance increase (middle).

Once water is injected to raise engine efficiency, the evaporation during the intake stroke leads to lower cylinder charge temperature, which in turn makes the engine more resistant to knocking combustion (the values marked with * in figure 1.2 refer to wide open throttle at 2000 rpm). Keeping the same engine configuration, this allows advancing the spark time, moving the combustion process towards early phasing. This regulation has two effects; on one side the indicating efficiency steps up, on the other side, the exhaust gas temperature decreases. The lowering of T_3 becomes noteworthy to extend the lambda one operating points through the all engine map if one wants to maximize as well the efficiency of the three-way catalyst and therefore reducing emissions (the values marked with ** in figure 1.2 refer to wide open throttle at 4500 rpm). Changing target and considering performance-enhancing, water injection can be employed to keep constant fuel consumption while raising torque by exploiting the cooling power of its evaporation during the intake phase. If it is the case, lowering intake air temperature can be used to achieve higher volumetric efficiency and thus allowing the engine boosting its power density. As background for both implementation purposes, the dilution effect of water must not be neglected since the amount of the injected water is significant, influencing T_3, laminar flame speed and thus combustion efficiency (right side of figure 1.2). The possible gain in terms of fuel consumption with water injection ranges from 4 % to 8 % respectively considering WLTC or higher power-demand cycle such as RTS95 (the values marked with *** in figure 1.2 are calculated considering a vehicle of 1700 kg [26]).

1.2.2 Water physical properties

The purpose of the 3D-CFD analysis is the comprehension of water impact during the different phases of the engine cycle. About figure 1.3, the main thermodynamic properties of water are compared with the ones of gasoline [54] for temperatures relevant during the compression stroke. Water presents a six times higher heat of vaporization (HOV) compared to gasoline. The lower the temperature at which the evaporation starts, the higher the cooling effect. Water tends to evaporate less than gasoline up to 80 °C, then the trend reverses, once observing the pressure of saturation. Furthermore, water has a 30 % greater density and three times higher surface tension with respect to gasoline, which delays the breakup of water droplets. Finally, the greater specific heat capacity plays a role for the water acting as inert mass and absorbing heat, even though this property becomes less effective for charge cooling, considering the lower amount of water injected compared to gasoline in realistic water to fuel ratio concepts. The parameter *Water to fuel ratio* is commonly used to define the amount of injected water per engine cycle, and it is usually calculated as reported in eq. 1.1.

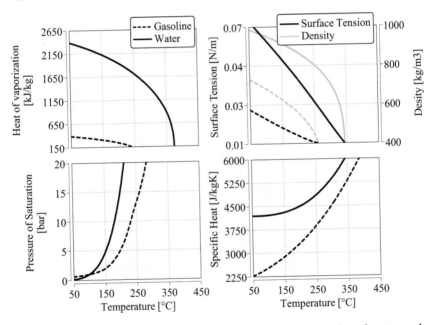

Figure 1.3: Comparison of the main thermodynamic properties of water and gasoline.

$$\frac{W}{F} = \frac{Massflow_{Water}}{Massflow_{Fuel}} 100 \qquad \text{eq. 1.1}$$

1.2.3 Limiting factors for water cooling effect

Other publications assess the unlikelihood of exploiting the theoretical charge cooling effect of water during evaporation [28] [7], identifying the heat loss towards the cylinder wall as the main reason for this. Effectively, the efficiency of the charge cooling up to ignition point (IP) is quite far from the theoretical contribution. In order to analyse water behaviour during the compression phase, three different 3D-CFD simulations were investigated, as reported in table 1.2.

The considered load point is 4500 rpm and 20 bar IMEP, corresponding to rated power for the engine at the test bench (in chapters 5, 6 and 8 this operating point will be identified as OP3). The engine at this operating point features 52 mg/cycle fuel running at $\lambda = 0.85$, meaning 7 mg/cycle needed for component protection (this engine does not need rich mixture for power increase and faster combustion to achieve the desired performances), while the amount of injected water was 14 mg/cycle, corresponding to 30 % water to fuel ratio. Fuel was injected at 320 °CA before firing top dead center (b.FTDC), while water at 390 °CA b.FTDC. Again concerning table 1.2, *Simulation 1* corresponds to fuel enrichment case, while simulations 2 and 3 were run at stoichiometric conditions with IWI at 15 bar water injection pressure. Particularly, *Simulation 3* used the HOV of gasoline instead of the one of water to describe water in the thermodynamic equations. All other liquid properties correctly corresponded to the one of water (surface tension, pressure of saturation, viscosity and density).

Table 1.2: Simulation settings to study in-cylinder temperature with fuel enrichment and with IWI

Sim [-]	λ [-]	W/F ratio [%]	Water inj. pressure [bar]	Water liquid properties [-]
1	0.85	0	-	Standard
2	1	30	15	Standard
3	1	30	15	HOV Gasoline

The cooling effect of additional fuel evaporation (*Simulation 1*) decreases in-cylinder temperature more than water injection (*Simulation 2*). Even if fuel is injected later in comparison to water, its evaporation tendency is higher and the temperature reduction is already visible at 135 °CA b.TDC. Water started evaporating later, producing a clear effect only after 80 °CA b.FTDC and covering the gap anyhow at the end of the compression stroke. At ignition point, the amount of evaporated water is 16 % to the total injected water, while fuel evaporation is almost completed (98 % to the total injected fuel). The temperature reduction due to additional fuel evaporation appears comparable to the one of water, even if the fuel surplus needed for component protection is the half to water mass. The low amount of evaporated water mass is the reason for the low cooling effect of injected water. The effect of higher HOV of water could be seen if water evaporation were comparable with the gasoline one.

On the other hand, locally the effect of high HOV let the surrounding fluid close to evaporated zones facing a huge temperature reduction, preventing, as a result, the evaporation of further water droplets, due to the decrease of the pressure of saturation. This combined effect slows down water evaporation and limits the cooling power of the higher HOV of water, as depicted in figure 1.4. Furthermore, by evaporating, the decrease of the charge temperature causes a growth of the polytrophic coefficient and a decrease of mixture specific heat capacity, which brings instead heating the mixture. However, these contributions are about 1-2 % compared to the temperature reduction due to the phase change (HOV effect). As a test, *Simulation 3* was considered where water HOV is set equal to the one of gasoline. In this case, the water evaporation rate at ignition point reaches 40 % compared to 16 % of the real case. These results confirm again the evidence that high HOV of water inhibits its evaporation.

Going further figure 1.5 quantifies the loss of temperature reduction for each simulation case. As reported in eq. 1.2, the calculations show the ideal temperature reduction (ΔT_{Ideal}), considering both for gasoline and water, the heat balance under full evaporation condition. These curves represent the mass evaporating instantaneously during the intake stroke for different pressure and temperature values (and relative variable specific heat, HOV and pressure of saturation). Under this hypothesis, water can overcome fuel in reducing temperature (43 °C vs 28 °C as peak reduction respectively) as shown through dashed lines in figure 1.5.

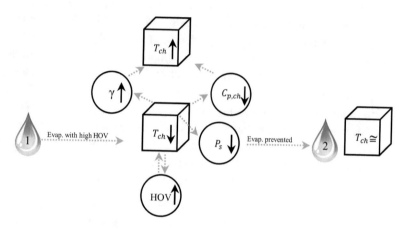

Figure 1.4: Effect of local charge cooling due to high HOV and resulting changes in the charge thermodynamic properties.

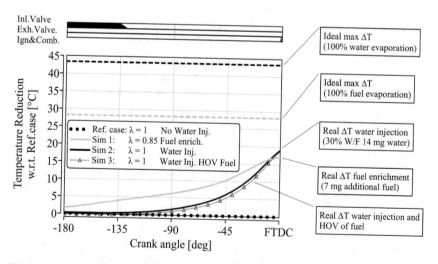

Figure 1.5: Temperature reduction of fuel enrichment case, IWI at $\lambda = 1$ case and IWI at $\lambda = 1$ with water property but HOV of fuel.

$$\Delta T_{Ideal} = \frac{\frac{W}{F} m_a HOV(T)_w + Q_{wall}}{\lambda \, \alpha_{st} c_{p,charge}(T)} \qquad \text{eq. 1.2}$$

where

ΔT_{Ideal}	Temperature reduction if all water injected mass evaporates
$\frac{W}{F}$	Water to fuel ratio
m_a	Induced air mass
$HOV(T)_w$	HOV of water (function of charge temperature)
T_w	Water temperature
Q_{wall}	Heat transfer through the wall
λ	Air excess
$c_{p,charge}$	Charge specific heat

Once the real evaporated mass is taken into account using eq. 1.3 (ones for fuel and once for water), fuel and water have the same effect for temperature control up to -80 °CA. From -80 °CA on, the higher in-cylinder temperature leads to a decrease of HOV of water and at the same time, to higher pressure of saturation (with reference back to figure 1.3), allowing water to evaporate and exploiting its cooling effect on the charge finally.

$$\Delta T_{Ideal,realevaporatedmass} = \frac{m_{w,vap} HOV(T)_w + Q_{wall}}{m_{charge} c_{p,charge}(T)} \qquad \text{eq. 1.3}$$

By computing the heat losses through the wall of the system, *Simulation 2* (with water injection) shows a negligible reduction of the heat transfer to *Simulation 1*, probably due to the slight decrease of the charge temperature leading to lower temperature gradient towards the wall. Still considering figure 1.5, the continuous curves describe the effective temperature reduction to reference case ($\lambda = 1$ and no water injection), both for the fuel enrichment and for the water injection and $\lambda = 1$ cases.

$\Delta T_{Ideal,realevaporatedmass}$	ΔT_{Ideal} considering real evaporated mass
$m_{w,vap}$	Evaporated water mass
m_{charge}	Mass of the charge

The power of water injection pointed out clear advantages with respect to fuel enrichment when investigating in-cylinder peak temperature and T_3. Referring to figure 1.6, the plot highlights in-cylinder temperature reduction (continuous lines) and in the exhaust runner (dashed lines) in relation to the reference case, respectively with fuel enrichment (grey lines) and water injection (black lines). The effect of water as inert mass ($14 mg/cyc$ water vs $7 mg/cyc$ additional fuel) and on the flame front propagation lowers the combustion peak temperature up until 140 °C, decreasing pressure gradient and probably NOx formation much more to the case at $\lambda = 0.85$. With water injection, the global temperature chain reduces up to the exhaust gas temperature, where water features an additional cooling of 20 °C compared to the fuel enrichment case. Water vapour is not only influencing combustion development, but the liquid mass still present at ignition point induces a post-evaporation during the next engine phases, leading to a further temperature decrease to fuel enrichment, as explained later in chapter 8.

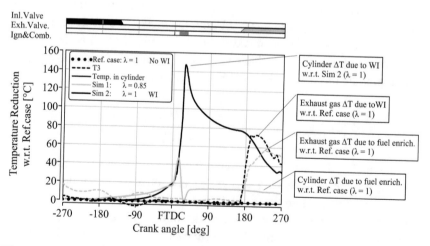

Figure 1.6: T_3 and in-cylinder temperature reduction with fuel enrichment and with IWI ($\lambda = 1$), with respect to reference case (no water injection and $\lambda = 1$).

The outcome of this analysis was used to improve the aforementioned indirect injection approach and to select the most promising concepts between indirect and direct injection strategy, by maximizing water evaporation and thus enhancing the charge cooling effect. Therefore a good injection strategy can be identified with the one increasing the water evaporation, but turning down the wall wetting (reduction of the liquid water at the liner).

1.2.4 Influence of ambient air humidity on water injection strategy

Ambient air flowing through the engine, already has certain water vapour content, depending on humidity rate of air. Considering the single-cylinder engine described in the next chapters 5, 6 and 8, it is useful to quantify the share of water vapour already present in the intake air. If this engine runs at 2500 rpm and 15 bar IMEP (this operating point will be later identified as OP2), it features 430 mg/cyc of air consumption. If the initial relative humidity of air (RH) is 70 % (average 2018 in Stuttgart, Germany [1]), almost 5 mg/cyc of water vapour are already present at ignition point without external water injection. At this load point injecting 30 % W/F ratio means that the additional water to be injected is 9 mg/cyc; thus the water content of air (humidity) is a meaningful contribution for the water mass balance into the engine. The water amount from humidity is of course already present into the system as gaseous phase, meaning that it has no leverage in reducing temperature through evaporation, acting finally just as additional inert mass and modifying the combustion process.

However, the amount of water vapour already existing in the air (humidity) influences significantly the maximum amount of external injected water when trying to reduce condensation. Especially when water is injected in the airbox, or well before the intake runner, there might be an unequal distribution of water droplets in the different cylinders since the local overrun of the saturation point. This means that the behaviour of one cylinder may differ one to each other, in terms of combustion phase and Knock occurrence, if the saturation point is overcome [24].

Therefore, the variation of saturation point, with respect to intake manifold conditions and initial relative humidity of air have to be correlated, in order to find an injection strategy that leads to a reduction of water condensation in the intake system. For this purpose figure 1.7 presents the maximum amount of water that can be injected preventing condensation, considering a transformation from initial *State 1* to *State 2*. *State 1* represents ambient conditions with different initial humidity contents, from neglecting humidity (1 %) up to 90 %, while *State 2* points 1.5 bar pressure (mid boost pressure for conventional DISI-Engines) and different intake manifold temperature conditions. The saturation point is strictly dependent on temperature changes, thus for different temperatures, the humidity of *State 2* is reported on the x-axis of figure 1.7. Table 1.3 helps reading figure 1.7 once intake manifold conditions of OP2 are examined.

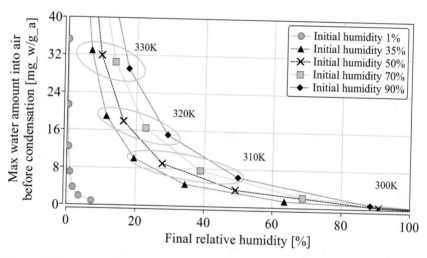

Figure 1.7: Maximum amount of water contained in the air before condensing, at the intake manifold pressure (1.5 bar) and at different intake manifold temperature conditions, for different initial humidity value of ambient air.

According to this, *State 2* now characterizes for 1.5 bar pressure and 310 K temperature. The first column of table 1.3 lists different humidity rate of ambient air (*State 1*), while the second column records the respective humidity rate after transformation from *State 1* to *State 2*. Then the third column reports the maximum water mass that can be injected, starting from the humidity rate of each *State 2*, up to 100 % RH, when reaching the saturation point (y-axis of figure 1.7). For injection strategy calibration is more convenient to convert the water vapour amount that can be handled by intake manifold air, in the respectively W/F ratio according to the fuel amount needed to run OP2 (30 mg/cyc fuel, at $\lambda = 1$). In the fourth column it can be seen that at 70 % initial relative humidity, at intake manifold conditions, the humidity of air reaches 39 % which in turn limits the maximum water to fuel ratio before condensation at 11 %.

What is again interesting to highlight is the delta between the maximum W/F ratio before condensation and the case by neglecting initial air humidity. The fifth column reveals that 90 % initial relative humidity of air reduces the possibility of the injected water evaporating and thus condensing up to 50% with respect to neglect initial relative humidity. This analysis represents the first step towards the calibration of the water injection strategy considering the environmental conditions.

Table 1.3: Psychrometric properties of intake manifold air from *State 1* after the transformation to *State 2*

RH1 [%]	RH2 [%]	Max water vapour in the air before condensation [mg_{water}/g_{air}]	W/F ratio [%]	Delta compared to no humidity case [%]
1	0.6	12	17	0
35	19	10	14	-17
50	28	9	13	-25
70	39	8	11	-33
90	50	6	9	-50

Furthermore, the optimization of injection strategy must be evaluated as well concerning the engine cycle, once evaluating water injection timing. When selecting the right timing, together with the saturation point, is useful to convert the previous analysis into the map presented in figure 1.8. Here, different combinations of charge temperature and pressure are grouped, generating a

saturation map, where z-axis displays the maximum amount of external water to be injected before condensation (already in terms of water to fuel ratio).
The saturation map calculates the saturation point considering 70% initial relative humidity of the air. Additionally includes to the plot, the temperature
and pressure combinations of the intake manifold (black points in figure 1.8)
for different injection timing.

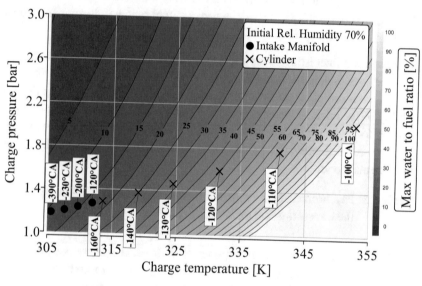

Figure 1.8: Saturation map considering different water to fuel ratio WI strategies such as IWI (black points) and DWI (crossed points), for 70
% initial RH.

On the other hand, the crossed points represent pressure and temperature combinations realized in the cylinder during the compression stroke and therefore,
they correspond to the flow field that water would face once it is directly injected in the cylinder (DWI). For early injection timing, the temperature conditions limit the maximum amount of water vapour held by air. Critical are the
circumstances in the intake manifold, where one can inject up to 17% W/F
ratio to avoid condensation, and for very early timing this reduces up to 10%.
The in-cylinder flow field allows a more extensive water range to be considered.

For example, if 30% W/F is needed, by imposing a start of water injection later than 130°CA b.FTDC prevents condensation problem (still considering 70% initial humidity).

The map indicates the maximum water amount that can evaporate in the air, considering the specific initial humidity. Nevertheless, the quantity of water able to vaporize depends on the working fluid local conditions and mainly, the vaporization takes place when the vapour pressure remains below the saturation pressure, which is in turn strictly related to the mixture temperature, as shown in figure 1.3.

By injecting later than 100°CA b.FTDC directly in the cylinder, theoretically no limitation exists in terms of the possible amount of evaporated water mass. In reality, the local conditions of the mixture, the time available for its formation, its inhomogeneous distribution do not allow the water to fully evaporate at ignition point, as previously discussed and as it will be shown in chapters 6 and 8. Simulations pointed out that the contribution of humidity does not affect so much the combustion process (particularly the flame speed) since the water is already vaporized into the charge. Nevertheless, considering the initial humidity content plays an essential role in the maximum amount of water that can vaporize.

1.3 The role of 3D-CFD virtual engine development

Three-dimensional (3D) CFD simulations allow the most detailed temporal and spatial resolution of flow-influenced processes inside internal combustion engines (ICEs) [35]. This possibility is essential when investigating the gas exchange (mainly influenced by the channels and the combustion chamber geometry), fuel injection and mixture formation (spray propagation, charge homogenization, local enrichment and wall impingement), combustion (significantly influenced by the induced turbulence) and emission behaviour. Since the higher complexity of nowadays engines, 3D-CFD simulations become a fundamental tool to study the interaction between the different engine parameters and to understand phenomena which could be controlled and exploited for improving efficiency. Besides, the dynamic of the market imposes a very high development rate, and thus the support of the virtual engine development could speed up the realization of a competitive engine. Simulations can predict the influence of engine geometry and control strategies on the engine performance over a wide range of operating points. Significant advantages over test bench investigations can be seen in a better reproducibility of analyses and for time and costs saving.

On the other side, the computational resources in terms of the required hardware and the related CPU-time are prohibitive with too fine mesh discretization. Consequently, the 3D-CFD-domain has also to be limited to the combustion chamber. This practice introduces boundary conditions that have to reproduce the part of the engine missing in the 3D-CFD-domain. The setting of the boundary conditions, e.g., due to the difficulty in determining or measuring the flow field in these regions can be very often a source of inaccuracy that irremediably compromises the overall quality of the results and drastically reduces the benefits of such resource investments [11].

Despite the enormous potential of 3D-CFD-simulations, their implementations remain quite limited in comparison to other simulation tools. The development of the 3D-CFD-tool QuickSim introduces a new approach in the three-dimensional analysis of internal combustion engines [11]. This tool tries to take advantage of the potentiality of the traditional 3D-CFD-approach combined with a remarkable reduction of the drawbacks mentioned above, ensuring a significant contribution in the engine development process. The future reduction of costs for computational resources and their increasing performance will surely support this task.

1.4 Future of ICEs and electro mobility

It is nowadays common thought that electric cars are the solution for reducing emissions and accomplishing cleaner air. In contrast, a German study [9] shows that a *Tesla* is no less polluting than a *Mercedes* equipped with a diesel engine. At first, the championship of the most eco-sustainable vehicle is played in a world where energy production has costs, in terms of pollution, extremely different from country to country. Therefore, the comparison of the environmental sustainability of an electric vehicle against a combustion vehicle depends on the considered country. It is not possible to take into account only CO_2 produced by a vehicle in use (*Tank to wheel approach*), equal to about zero for an electric vehicle. Instead, it is necessary to calculate the pollution produced during the entire life cycle of an electric car, by taking into account the cost in terms of CO_2 for electricity production generated to power the car and to dispose of the batteries (*Well to wheel approach*). It is for sure true that combustion engine produces, in addition to CO_2, oxides of nitrogen (*NOx*), unburned hydrocarbons (*HC*) and carbon monoxide (*CO*), pollutants linked to the combustion process. However, the post-treatment of these pollutants in modern diesel and gasoline engines (Euro 6d) realizes with a high conversion efficiency. These pollutants are certainly not produced by an electric car while driving, but since they are combustion products, they are as well generated when producing electricity through fossil fuel. Unfortunately, there are no precise data to quantify the contribution of these pollutants, linked to the production of electricity at the moment. An unambiguous comparison today is not possible since the exhaust systems of power plants discharge at several meters in height and are not subject to the same strict constraints in terms of post-treatment of exhaust gases, as for cars. Instead, they can be considered as a percentage of the CO_2 produced, but without exhaust gas treatment as in cars. Such a comparison would, therefore, be definitely unfavourable to an electric car if the energy used to power it would be produced from fossil fuels. The above mentioned German study has, in fact, only attributed to the energy used by an electric car a share of CO_2 according to the country of energy production. It turns out that, by using fossil fuel to produce energy, the share of CO_2 "emitted" by an electric car is comparable to that of the combustion of a diesel engine, if considering the additional CO_2 generated by battery disposal processes. Comparing diesel to petrol engines, it appears completely not reasonable the denigration campaign on diesel engine recently pursued and more specific, unjustified on a scientific level. The diesel engine has a higher thermodynamic efficiency than a gasoline one and emits on average 15 % less CO_2 at the same distance covered. In addition, all Euro 6 diesel engines are equipped with a particulate filter to reduce particulate matter, a device present on very few Euro 6 petrol cars. For those who study the emissions of a car, it is clear that a Euro 6 Diesel emits less particulate matter than a Euro 6 petrol without a particulate filter. So if the problem is to reduce particulate matter, not only blocking a Euro 6 diesel does not make sense, but it would be insane to encourage the use of a Euro 6 petrol.

1.4.1 Comparison between BEVs and ICEs

It is common practice today to wonder whether or not internal combustion engines would have a future. The real question is how it would be possible to refine further a technology that was born almost 200 years ago. At the moment, forecasts agree that purely electric vehicles (BEV) will account for 20 % of all vehicles on the market by 2030 in the best scenario [2]. Following this forecasts means that the internal combustion engine will play a significant role for a long time to come. The reasons are economic, technological, infrastructural. Let us take a practical example with a high-volume car like the *Peugeot 208*. The new *Peugeot 208* is offered with petrol, diesel or purely electric engine. The diesel version costs 18000 € while the electric one starts from 37000 € . The 208 is a city-car, which means that to be sold must have a competitive price, while the electric version costs twice the diesel version.

The electric version is a bit more performing, however, its range is 340 km on WLTC cycles, while on the same cycle the diesel version with 40-litre tank completes about 1000 km with a full tank. Then the supply problem must be considered. The manufacturer guarantees that 80 % of the battery can be recharged in 30 minutes with a 100 kWh column. However, the most common ones are 22 kWh, while the domestic network is 3.3 kWh (Italy). Furthermore, the electric vehicle must be recharged at home for 25 hours (3.3 kWh) or for 5 hours with a 22 kWh column to accomplished 340 km. In the case of a public charging station, then it must be bear in mind how many cars they charge at the same time. With two cars charging, the charging time of the single-car doubles and generally the single-car charging time grows proportionally.

The energy costs depend on the country concerned and on the speed of charging. In Italy, for example, it ranges from about 0.3 € /kWh for a 22 kWh charging station to 0.55 € /kWh for a fast-charging station of 100 kWh. So to cover about 1000 km, which with the diesel version could be made with a full tank and about 60 € , with the electric version two recharges are needed of one hour with 100kWh column and a cost of about 160 €. Therefore, it is straightforward to understand that electric mobility, as nowadays conceived, it can be considered acceptable only in countries with a few million inhabitants, with a high per-capita income and not without massive infrastructure investment.

1.4.2 Alternative fuels

As a possible solution for eco-sustainable mobility, but at the same time as it is feasible, the parallel development of two concepts should be pursued. The first is to make current petrol and diesel engines even more efficient (the margins are considerable, and one of the most promising technology is water injection). The second is the development of synthetic fuels. Synthetic fuels, and especially e-fuels (synthetic fuels produced through renewable energy), are gasoline or diesel fuels created in the laboratory by binding carbon and hydrogen. Again, nothing exceptional so far. Traditional fuels consist of carbon and hydrogen chains. The particularity lies in the sources where these two components can be found. Carbon is an element that constitutes CO_2 and therefore the idea is to use CO_2 from specific combustion processes that cannot be replaced (CO_2 from industrial ovens for food, plastic, concrete) and to bind it with hydrogen produced by electrolysis from water, using electrical energy derived from renewable sources. By the time being, this is the only way towards sustainable and $CO_2 - neutral$ mobility as they say in jargon. This process is commonly indicated as CO_2 neutral because the CO_2 generated into the atmosphere by burning fuel in a car engine is absorbed in the production process of the gasoline itself. Moreover, these fuels have an extremely controlled chemical composition and generally fuel-design which, when burned, reduce emissions and increase engine efficiency.

2 Fundamentals of CFD for ICEs

The majority of current ICEs are no longer naturally aspirated, instead they employ boost systems to increase the air density. Usually, this is realizes via a turbomachine and particularly with a turbocharger, which exploits exhaust gas enthalpy to compress fresh air and to raise the engine volumetric efficiency. The objective is to suck as much air as possible decreasing engine displacements while keeping the same rated power. This practice is generally known as *engine downsizing*. Additionally, the engine speed can be lowered to reduce friction (downspeeding)[22].

Commonly the engine power can be obtained through eq. 2.1. With the effective power P_e , the factor i for two or four stroke engines, the break mean effective pressure (*BMEP* or *pme*) and the engine displacement V_{cyl}.

$$P_e = i * n * pme * V_{cyl} \qquad \text{eq. 2.1}$$

Since in a downsized engine, V_{cyl} is smaller, *pme* has to be increased in order to maintain the same performances. The smaller the engine displacement, the reduced the specific heat losses, the engine frictions and the share of the auxiliary system on the overall contribution to fuel consumption. On top of that, the implementation of a turbocharger for gasoline engine imposes a limit on the maximum exhaust gas temperature before the turbine, considering the mechanical material limit that the turbine can afford. Therefore, a standard solution is to run the engine with a rich mixture in order to exploit the enthalpy of evaporation of additional fuel and at the same time the presence of higher inert mass to cool down the exhaust system. The second cooling effect is much lower that the first mentioned one, since the modest additional mass with respect to the global charge. On the other hand, diesel engines do not require fuel enrichment because the global lean mixture of such combustion produces already lower exhaust gas temperature. Also, in diesel engines, the adoption for higher compression ratio (no limitation in terms of Knock onset) allows realizing higher expansion ratio and therefore cooler temperature at the end

of combustion. Unfortunately, the drawback of the enhanced power density
of a downsized engine is represented by design challenges requiring advanced
solutions [29]. In order to support the increased complexity of future engine
development, three main simulation tools are combined in the so-called *Virtual
engine development*:

- Real working process calculation
- One-dimensional (1D) CFD simulation
- Three-dimensional (3D) CFD simulation

The first consists of a 0D approach where the engine operating cycle analysis
is based on the energy conservation equations, but no temperatures and concentration distributions are considered.

One-dimensional fluid-dynamics concentrates on the thermodynamic analysis
of the cylinder process by combining simplified fluid dynamic simulation for
all pipes (intake and exhaust system).

Finally, the 3D-CFD simulation is a general flow field analysis where real engine geometries are discretized in defined volumes, and the conservation equations are solved for each of them.

2.1 The real working process analysis

The real working process is the first step towards the investigation of engine
thermodynamics processes. It is based on the solution of the energy conservation equation (eq. 2.2), the mass conservation equation (eq. 2.4) and thermal
equation state (eq. 2.5) reported below.

- **First law of thermodynamics**
  Given the principle of energy conservation the sum of the energy transfer
  in a control volume, such as the engine cylinder can be written as in eq.
  2.2.

$$Q_B + Q_W + H_I + H_E + H_L + W = 0 \qquad \text{eq. 2.2}$$

where Q_B is the fuel heat release energy, Q_W the wall heat transfer, H_I the intake enthalpy, H_E the exhaust enthalpy, H_L the leakage enthalpy W the work produced on the piston, considering the whole engine cycle. The differential notation of eq. 2.2 describes the change of internal energy U over time and allows a crank angle-resolved analysis of the engine operating cycle, as written in eq. 2.3.

$$\frac{\mathrm{d}Q_B}{\mathrm{d}\varphi} + \frac{\mathrm{d}Q_W}{\mathrm{d}\varphi} + \frac{\mathrm{d}H_I}{\mathrm{d}\varphi} + \frac{\mathrm{d}H_E}{\mathrm{d}\varphi} + \frac{\mathrm{d}W}{\mathrm{d}\varphi} + \frac{\mathrm{d}H_L}{\mathrm{d}\varphi} = \frac{\mathrm{d}U}{\mathrm{d}\varphi} \qquad \text{eq. 2.3}$$

- **General mass balance**

In addition for the mass conservation a mass balance can be written as in eq. 2.4, where the change of cylinder mass m_C results from outflow of exhaust gas m_E and leakage mass m_L due to blow-by as well as inflow of fresh load m_I and fuel mass m_B in case of direct injection engines.

$$\frac{m_C}{\varphi} = \frac{m_I}{\varphi} + \frac{m_E}{\varphi} + \frac{m_L}{\varphi} + [\frac{m_B}{\varphi}]_{DI} \qquad \text{eq. 2.4}$$

- **Thermal equation of state**

$$p \cdot V = m \cdot R_s \cdot T \qquad \text{eq. 2.5}$$

describing the relation between the thermal state variables of an ideal gas, i.e. pressure p, volume V, mass m, specific gas constant R_s and temperature T.

2.2 1D-CFD simulation

In addition to the *Real Working Process Analysis*, the 1D-CFD simulation allows a spatial resolution of the engine along the main flow direction. The simulation domain is then divided into a finite number of sub-volumes, and the flow field is calculated through conservation equations (continuity, momentum and energy) [11]. The combustion process is modelled through phenomenological or quasi-dimensional correlations, which consider influencing factors like the combustion chamber geometry or the in-cylinder flow motion (Tumble, Swirl).

These combustion models are suitable to calculate changes in the operating conditions (e.g. engine speed, lambda, EGR). For this reason they can be considered as predictive models, in opposition to pure empirical ones. Engine components with more complex geometries (e.g. turbocharger) are usually implemented via characteristic maps [11].

Although 1D-CFD models allow a rapid definition of many engine parameters and general design features, they have limited application when the goal of the analysis is an assessment on the geometry of the engine and especially on injection development, which is an essential aspect when for example water injection is considered. In addition, this approach shows its limit when assessing geometry influences on the flow field, e.g. of the intake and exhaust channels, different combustion chamber or injector geometries. These limitations can be compensated by adopting 3D-CFD simulations.

2.3 3D-CFD simulation

2.3.1 Fundamental equations

3D-CFD simulations represent the most detailed and comprehensive approach to numerically investigate fluid dynamical problems. The methodology consists of the discretization of the engine parts into a computational grid. In each cell of the grid, the reactive flow field is resolved. The description of the fluid flows is accomplished by coupling the conservation equations for mass, momentum and energy with an equation of state. An overview of the main equations is resumed in the current section [4, 10, 11, 55].

It is convenient to derive from the Euler formulation the conservation equations for momentum, energy and mass following eq. 2.6.

$$\frac{\partial f}{\partial t} + div\left(\vec{\Phi}_f\right) = s_f + c_f \qquad \text{eq. 2.6}$$

The Euler formulation states that a change in the variable density $\partial f / \partial t$ can be caused by a flux through the volume surface $\vec{\Phi}_f$, by a source or sink (s_f) or a long-range process (c_f) like radiation or gravitation. Particularly, $f(\vec{x},t) = dF/dV$ represents the density function of the conservation quantity (mass, momentum or energy are extensive quantity). The current density $\vec{\Phi}_f$ quantifies the amount of quantity F, which flows through the surface area per time unit.

- **Mass conservation**
  From eq. 2.6 it can be obtained the mass conservation equation by replacing to the extensive variable $F(t)$, the mass density ρ. In addition the flux density can be written as the product of local flow velocity ($\vec{v}$) and the density itself (ρ). Both the source therm s_f and the long range one are equal to zero as mass cannot be created or destroyed by the transformation considered. Following this steps eq. 2.7 can be derived.

$$\frac{\partial \rho}{\partial t} + div\,(\rho \vec{v}) = 0 \qquad \text{eq. 2.7}$$

eq. 2.7 can be written for different species such as air and fuel vapour, by replacing the mass (m) with the mass fraction $w_i = m_i/m$ of species i. The mass density can be replaced by the product of local flow velocity v_i composed of the mean flow velocity $\vec{v}$ and the diffusion velocity $\vec{V}_i$. If one considers as $\vec{j}_i$ the derived mass flux created by the diffusion velocity, eq. 2.7 can be written as eq. 2.8.

$$\vec{\Phi}_f = \rho_i \vec{v}_i = \rho_i \left(\vec{v} + \vec{V}_i \right) = \rho_i \vec{v} + \vec{j}_i \qquad \text{eq. 2.8}$$

Since the combustion process may cause some species transformation (chemical reactions), a source term is required which can be written as the product of species molar mass M_i and species molar formation rate ω_i.

$$\frac{\partial \rho w_i}{\partial t} + div\,(\rho w_i \vec{v}) + div\,\vec{j}_i = M_i \omega_i \qquad \text{eq. 2.9}$$

- **Momentum conservation (Navier-Stokes equation)**
  When considering the conservation of momentum the density variable in eq. 2.6 is replaced by the momentum density ($\rho \vec{v}$). The flux term is now composed by a convective part $\rho \vec{v} \otimes \vec{v}$ and the second order stress tensor $\overline{\overline{P}}$ which takes into account the momentum change due to viscous effects $\overline{\overline{\Pi}}$ and the momentum change due to pressure $p\overline{I}$.

$$\vec{\Phi}_f = \rho\vec{v}\otimes\vec{v} + \overline{\overline{P}} = \rho\vec{v}\otimes\vec{v} + p\overline{\overline{I}} + \overline{\overline{\Pi}} \qquad \text{eq. 2.10}$$

In addition the momentum equation must consider gravitation as long range terms ($\rho\vec{g}$). For this reason eq. 2.10 can be written as eq. 2.11.

$$\frac{\partial(\rho\vec{v})}{\partial t} + div(\rho\vec{v}\otimes\vec{v}) + div\,\overline{\overline{\Pi}} - grad\,p = \rho\vec{g} \qquad \text{eq. 2.11}$$

- **Energy conservation**

In the conservation of energy the sum of the internal energy u, the kinetic energy $\frac{1}{2}|\vec{v}|^2$, the potential gravitational energy G and the heat of formation of the considered species h_f is equal to the total energy density ρe as reported in eq. 2.12.

$$\rho\left(u + \frac{1}{2}|\vec{v}|^2 + G + h_f\right) = \rho e \qquad \text{eq. 2.12}$$

The energy flux represents the sum of a convective term $\rho e\vec{v}$, the energy transport due to pressure and shear stress $\overline{\overline{P}}\vec{v}$ and the energy transport $\vec{j}_q$ due to heat conduction. The source term s_f since the assumption of energy conservation can be set to zero. Radiation effect (q_r) can be considered in the long-range term c_f. Finally, The thermo-chemical enthalpy (h_{tc}) can be replaced by the sum of the thermal enthalpy h and the heat of formation h_f. When considering these assumptions, the energy equation assumes the form shown by eq. 2.13.

$$\frac{\partial(\rho h_{tc})}{\partial t} - \frac{\partial p}{\partial t} + div\left(\rho h_{tc}\vec{v} + \vec{j}_q\right) + \overline{\overline{P}} : grad(\vec{v}) - div(p\vec{v}) = q_r \quad \text{eq. 2.13}$$

given the relation between the specific internal energy and the specific enthalpy resumed in eq. 2.14.

$$h = u + pv \qquad \text{eq. 2.14}$$

2.3.2 Turbulence modelling

Two different regimes are usually identified in the model of flow processes, the laminar regime in which the flow is regularly dispersed and the turbulent regime in which random non-stationary structures of different size exist. The shear stresses, in particular, produce vortices of different size and shape. They are continuously drawing energy from the main flow while shedding into an increasing number of smaller vertices [4, 29]. During this vortex cascade, energy is dissipated back to the main flow. As usually convenient to describe phenomena dependent from copious parameters, a non-dimensional number can be used to describe the different regimes. Notably, the *Reynolds number* represents the ratio of inertial to viscous forces as shown in eq. 2.15.

$$Re = \frac{\rho |\vec{v}| l}{\mu} \qquad \text{eq. 2.15}$$

where the inertial forces are considered as the product of density (ρ), velocity ($|\vec{v}|$) and a characteristic length (l), while the viscous forces are represented by the dynamic viscosity (μ). Depending on the application, a specific value of the *Reynolds number* describes the transition between laminar and turbulent flow.

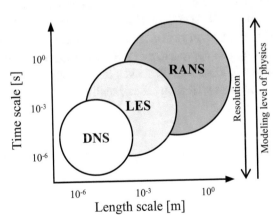

Figure 2.1: Modeling approaches for the description of a turbulent flow field [55].

The flow in an internal combustion engine is definitely turbulent; therefore, the model of turbulence is essential for the accuracy of 3D-CFD simulations. In order to solve the conservation equations described above, different turbulence model approaches are available. In figure 2.1, these approaches are presented in relation to time and length scale, but as well to their resolution and physic complexity [55].

The most detailed methodology is the *Direct numerical simulation* (DNS) which allows capturing even the smallest turbulent length scales. Here, the equations are directly and numerically solved avoiding the use of a turbulence model completely. Since the high computational effort, DNS approach is not suitable for the simulation of engine fluid dynamics, especially with high Reynolds numbers [3, 8, 11].

The *Large eddy simulations* (LES) approach considers grid models to reproduce small-scale turbulence components, but large eddies are still explicitly calculated, requiring as well highly refined computational meshes with millions of cells [11].

The standard for the simulation of engine flow fields and, in particular for the purpose of virtual development is, therefore, the application of *Reynolds Averaged Navier Stokes* (RANS) equations [3]. Here, models describe the turbulent flow entirely, and an explicit numerical solution of the equations is omitted. Considering this approach, a significant reduction of computational effort in comparison to DNS or LES makes RANS rather applicable to large computational domains as well as in case of high Reynolds numbers [55].

3 The 3D-CFD Virtual Test Bench

This chapter provides an overview of the CFD methodology used in the current work. Starting from the equations presented in chapter 2, the classical CFD approach is extended towards a time-effective analysis dedicated expressly to the development of internal combustion engine (see also chapter 1.3). This analysis is marked out as *3D-CFD Virtual Test Bench* because of the massive amount of configurations tested through simulations and for the similarity to the experimental methodology usually adopted for the investigation of the engine thermodynamics at the test bench.

3.1 QuickSim approach and methodology

The numerical investigations within the presented project were carried out with the 3D-CFD tool QuickSim, which has been developed at Forschungsinstitut für Kraftfahrwesen und Fahrzeugmotoren Stuttgart (FKFS), in collaboration with Institut für Fahrzeugtechnik Stuttgart (IFS), former Institut für Verbrennungsmotoren und Kraftfahrwesen (IVK). Using the commercial software STAR-CD as a solver, QuickSim is a fast response 3D-CFD tool, which is specifically designed for the simulation of ICEs. By using ICE-adapted and improved computational models, coarser meshes compared to traditional 3D-CFD approaches can be used without sacrificing the quality of the simulation results. As a result of this, the time expense for a simulation is highly minimized (2 hours for an operating cycle of a full engine using a 12 cores CPU) [11]. This strategy, not only allows the extension of the simulation domain to a full engine, but also the calculation of several successive operating cycles (up to transients) in a reasonable time frame.

Figure 3.1 reports a comparison between the time needed for electronic control unit (ECU) processes (real time) and the one needed for the *Virtual engine development process* mentioned in chapter 2 (*Real working process analysis,*

1D-CFD simulations and *3D-CFD simulations*). Among these a further distinction is realized considering the time needed for simulations with traditional CFD approach with respect to the one necessary with QuickSim.

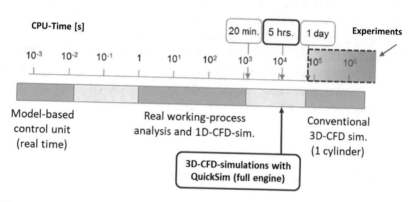

Figure 3.1: Time scale of simulations and process during engine development.

Among the different models used in QuickSim, of which a detailed description is provided in [11], table 3.1 resumes the main ones adopted or modified in the current work.

Table 3.1: Overview of the simulations characteristics in QuickSim

Simulation methodology	Multiphase RANS
Turbulence modeling	Two-Layer k-ε
Multi-phase flow	Euler / Lagrange
Combustion – Flame Propagation	Two-zone Weller-model adapted to GDI-combustion for spark ignition flame propagation and self-ignition model.
Solver	Coupled flow and energy with 2nd order implicit method
Time modeling	Stationary

3.2 Accurate fuel description

The combustion process of a conventional gasoline fuel involves more than 7000 chemical species. Due to such complexity, reduced chemical mechanisms are commonly implemented for the simulation of chemical reactions. The combustion is then described by simplified reaction schemes which include only a limited number of relevant species.

Due to the molecular complexity of ordinary fuels, experimental and computational investigations of fuel reaction kinetic during combustion is virtually impossible. Therefore, fuels are usually represented by surrogates of more straightforward molecular composition. As described in the literature [41] [31], surrogates are composed by some of the most representative species of the leading chemical families present in the fuel, like those listed in table 3.2.

Table 3.2: Most representative chemical species used for gasoline surrogates

n-alkanes	n-heptane, n-pentane
iso-alkanes	isooctane, isopentane
cyclo-alkanes	cyclohexane, cyclopentane
Aromatics	benzene, toluene
Olefins	2-methyl-2-butene, cyclopentene
Oxygenates	ethanol, ETBE

Different methodologies can be employed to determine surrogate composition. The most widely used surrogates for gasoline fuels are a mixture of n-heptane and isooctane, commonly called primary reference fuel (PRF), or a ternary mixture of PRF plus toluene (TRF). Experimental researches show that these simple compositions are suitable to reproduce with sufficient accuracy properties like the laminar flame speed of commercial fuels, but they show limitations in Knock prediction. These simple surrogates cannot accurately describe the remarkable influence of small concentrations of chemical species like olefins and oxygenate, which remarkably increase the Knock resistance of the fuel, while not significantly affecting their laminar flame speed. For these reasons, a more complex surrogate, including at least one species for each relevant chemical family present in the fuel, should be adopted. In general, two approaches can be used to simulate thermodynamic properties variation and chemical reactions that occur during combustion. One approach consists of implementing a sufficiently detailed mechanism (with more than 100 species) within the 3D-CFD simulation environment. As shown in [5, 43], this methodology can ensure sufficiently reliable results but with a considerable increase in CPU-time. Generally, the higher the number of species and reactions contained in a chemical mechanism, the higher the calculation time. However, too simplified mechanisms are not able to accurately describe the real fuel behaviour. Alternatively, to reduce the calculation effort, thermodynamic properties of the mixture and information on chemical reactions (such as laminar flame speed and auto-ignition delays) can be read from external databases which are prepared separately with dedicated software. Different works such as [30] use a similar approach. Furthermore, with this methodology, it is possible to solve chemical kinetics calculations with accurate mechanism, even more, detailed than those normally used within 3D-CFD simulations due to CPU time constrains. This latter approach is the one used in QuickSim.

3.3 Water injection implementation

The working fluid in QuickSim is described by few scalars which, in combination with look-up tables, thermodynamically represent fresh charge, EGR (burned gas of the previous cycle) and burned gas (generated after the combustion process of the current cycle). The base six scalars composing the working fluid are conveniently resumed in table 3.3 and a detailed description of each of them can be found in [11].

For the purpose of separating the exhaust gas in front of the flame front, with the one arising from the combustion, it was necessary to generate respectively two different scalars " Burned gas " and " EGR ". Also, the compositions of " Burned gas " and " EGR "were reproduced by identifying their reactivity with respect to the fresh charge. Therefore the concept of " λ_{EGR} " and of " λ_{Burned} " was introduced. This approach allowed in the modern development of ICEs to consider for example phenomena like post oxidation, where the interaction between exhaust gas not yet saturated ($\lambda_{EGR} < 1$ or $\lambda_{Burned} < 1$) with fresh air could generate secondary combustion.

Table 3.3: Description of the base six scalars defined in QuickSim [11].

Scalar	Description
$Air_{Unburned,j}$	Mass fraction of fresh air
$Fuel_{Unburned,j}$	Mass fraction of fresh vaporized fuel
$EGR_{Air,j}$	Mass fraction of air that has previously produced EGR (burned gas of the previous operating cycle)
$EGR_{Fuel,j}$	Mass fraction of vaporized fuel that has previously produced EGR (burned gas of the previous cycle
$Air_{Burned,j}$	Mass fraction of air that has previously produced burned gas
$Fuel_{Burned,j}$	Mass fraction of vaporized fuel that has previously produced burned gas

In addition, tables describe the combustion products where properties, such as thermal enthalpy of the gas in the burned zone, are calculated as a function of the flow thermal state (temperature, pressure and λ_{Burned}). These tables are the direct consequence of the fuel model and its molecular composition which determines the final gas thermal-properties, and for this reason, a detailed fuel reproduction is needed as explained in paragraph 3.2.

For the modelling of water effects into the engine, the original description of the working fluid, based on six scalars, was extended with the introduction of three new scalars:

1. " Water ";

2. " EGR Water "(water coming from previous cycles);

3. " Burned Water "(water taking part to the combustion process, vaporized and crossed over by the flame front);

Accordingly, to this methodology, the scalars are not directly related to single chemical species, but they are optimized to identify and describe the working fluid in its main groups of interest (fresh gas, EGR and burned gas) during all the phases of the engine operating cycle. By means of this approach, it was possible to quantify directly:

- the mass of water vaporized in the different engine parts (intake/exhaust manifold, cylinder),
- the mass of water vapor involved into the combustion process (slowing down the laminar flame speed, increasing the ignition delay, influencing the onset of Knock),
- the water vaporizing at the end of the combustion process (post evaporation not influencing for example the laminar flame speed),
- the mass of water recirculated due to scavenging processes, or still present in the engine from a previous cycle,
- the mass of water accumulated at the liner (pre-assessment of wall impingement).

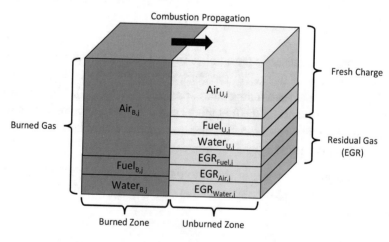

Figure 3.2: Partition of the scalars programmed in the j-cell of QuickSim, during combustion [49].

By coupling this working fluid definition, and the reproduction of real fuels, a new 3D-Knock model was also implemented [12], which was experimentally validated with water injection measurements. The modelling of all engine processes (in particular the combustion) was adapted and accordingly optimized for this formulation, as explained in [11]. The databases are prepared separately using Cantera, a suite of object-oriented software tools for problems involving chemical kinetics, thermodynamics and transport processes. As a result of this new fluid definition in the software, figure 3.2 schematically shows the new composition of a general 3D-cell during combustion coded in Quick-Sim. The scalars can be grouped into unburned ones (fresh scalars before being crossed by the flame front) and burned ones (already crossed by the flame front). With this approach, the flame front in each cell can be reproduced.

3.4 Auto-ignition model and Knock detection criterion

The fuel description is essential for the correct prediction of knocking combustion and therefore, a detailed model considering the fuel chemical composition was adopted [12, 13, 49]. Following this approach, the ignition delay, the onset of Knock and the laminar flame speed could be calculated as a function of the engine 3D flow field (pressure, temperature, composition, turbulence), but as well from the water content of the mixture. Since in QuickSim no detailed chemical reactions are directly implemented, a dedicated model describes the occurrence of auto-ignition. Most of the approaches developed in the last decade are generally based on the evaluation of an integral representing the pre-reaction state of the unburned mixture. The formulation reported in eq. 3.1 was initially proposed by Livengood and Wu [32]. They proved that the ignition delay time to auto-ignition of an air-fuel mixture for gasoline engines could be estimated by evaluating an integral representing the degree of chemical reaction progress and thus the pre-reaction state of the mixture.

$$\int_{t=0}^{t=t_e} \frac{1}{\tau_{ID}} \, dt = 1 \qquad \text{eq. 3.1}$$

In this equation, t is the elapsed time, t_e is the time at the end of the integration and τ_{ID} is the ignition delay to auto-ignition of the mixture at the current boundary conditions. Hence, if the ignition delay time τ_{ID} is known at every integration step, it is possible to predict when the auto-ignition of a mixture in a firing engine will occur (end of integration t_e). Ignition delay times τ_{ID}

can be either measured in a rapid compression machine or calculated using detailed kinetic reaction mechanisms.

The integral, according to its original formulation, is ordinarily calculated considering the average mixture conditions in the cylinder. However, such an approach has two main drawbacks: first, auto-ignition occurs due to the formation of radicals that depend on local charge conditions and significant approximations must be made if an average cylinder temperature is considered for the calculation of the integral. Secondly, phenomena like Knock in gasoline engines occur only if a certain mass ratio of mixture auto-ignites. Therefore, it is also essential to calculate the quantity of charge in auto-ignition conditions so that Knock occurrence can be correctly detected. A local integral counter has been implemented in QuickSim to overcome these problems, and its update realizes in every cell for each time step. With such a mean, it is possible to consider fluid mixing and to calculate the location and the quantity of charge in auto-ignition conditions [13].

At the test bench, Knock limit is found by advancing the ignition point until a previously defined Knock rate limit (expressed as a percentage of analyzed working cycles in which Knock is detected) is reached. Unfortunately, it is numerically tricky to reproduce pressure oscillations occurring at the test bench correctly and, therefore, a different approach is usually used to detect Knock in CFD-simulations. As experimentally established, an engine is in critical Knock conditions only if a sufficient amount of unburned mass auto-ignites. Thanks to the locally resolved auto-ignition model previously introduced, it is possible not only to identify the amount of mass in auto-ignition conditions but also its location at each time step. Besides, to establish a Knock criterion, it is necessary to find a critical value of mass in auto-ignition over which the operating point is considered to be in Knock. However, the mass in auto-ignition cannot be readily estimated experimentally, and it is not practically possible to have a direct comparison of simulation results with test bench measurements. Nevertheless, as shown in chapter 6, if a simulation reproducing a Knock-critic operating point is taken as reference (calibrated through measurement carried out at knocking limit conditions), the calculated mass in auto-ignition can be successfully used to compare different engine configurations and fuel types, predicting the Knock tendency.

3.5 Influence of water vapour on flame speed and auto-ignition

Vaporized water acts as an inert mass in the combustion process, which, similar to EGR, can significantly influence the combustion characteristics. According to reaction kinetics, an increase of both EGR and water concentration lead to an almost linear decrease in laminar flame speed and an almost linear increase in ignition delay time, as shown in figure 3.3.

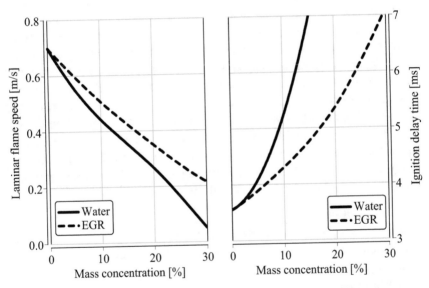

Figure 3.3: Influence of water and EGR concentration on laminar flame speed (left) and on ignition delay time (right) at 800 K, 80 bar and $\lambda = 1$.

Moreover, for the same mass concentration, the influence of water on laminar flame speed and ignition delay time is much stronger compared to EGR. This impact is mostly due to the different chemical properties of EGR to water. For the calculations, a surrogate composed by 44 % isooctane, 14 % n-heptane, 32 % toluene and 10 % ethanol was considered. The Lawrence Livermore National Laboratory developed the reaction mechanism used to solve the reaction kinetics in the current work, and it includes 324 species and 5739 reactions [33].

3.6 Combustion modelling

Once the simulation of reacting flows is considered a connection between the fluid's mechanical and chemical behaviour is required, and particularly, the source term in eq. 2.9 has to be defined. Reaction mechanisms for ICEs, which involve hydrocarbon combustion, are a highly complex system with thousands of species and hundreds of competing reactions. Moreover, the exact chemistry of the combustion processes and how one reaction step influences another is not well understood yet [11]. It is then somewhat reasonable to implement engine specific combustion models that focus on heat-release prediction, neglecting a detailed analysis of the combustion mechanisms which are not relevant in the prediction of the engine operating cycle. QuickSim follows this approach for the combustion model. In the software environment, four heat-release models are coded, and they are summarized below [11, 12]:

- **Spark-ignition**
  It represents a classical approach for the model of spark ignition engines (SI-Engines) combustion where the burn rate depends on the flame front propagation.

- **Self-ignition**
  It is a model which aims to reproduce self-ignition combustion such as HCCI, and the volume reaction of the cell manage the process.

- **Diffusive flame**
  This combustion approach supports the reproduction of the second part of diesel combustion, namely the diffusive flame propagation. This model switches together with the previous one, depending on the reaction rate, moving from self-ignition combustion to diffusive flame.

- **Post-oxidation of exhaust gases**
  It represents a similar model to the *Self-ignition* one, but the reactions develop mechanism is based on the amount of burned species concentration

like *CO* and the air available, considering as driven parameter exhaust temperature.

The approach used for the simulations in the current work is the *Spark-ignition* model since the investigated engine at the test bench was operated just with standard gasoline combustion. Nevertheless, the Knock model used for combustion optimization and seen in chapter 3.4 is based on a self-ignition model whose evaluation working principles are then same of the combustion self-ignition model mentioned before [12].

It is a common approach in SI-Engines to model the combustion flame as laminar flame speed which propagates through the unburned zone with a relative velocity S_L and when the flame front reaches a dimension comparable to the turbulent eddies it accelerates to the turbulent flame speed S_T. The turbulent flame front can be described as a laminar flame with an increased flame surface, as shown in figure 3.4.

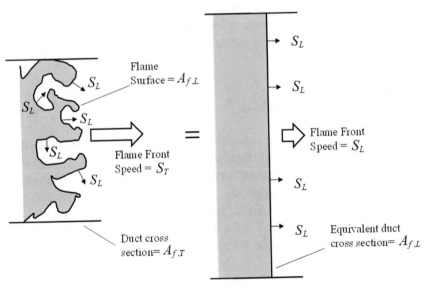

Figure 3.4: Schematic of turbulent flame propagation [11].

The link between laminar and turbulent flame speed is the wrinkling factor K, whose definition is reported in eq. 3.2.

$$K = \frac{A_{f,L}}{A_{f,T}} = \frac{S_T}{S_L}$$
eq. 3.2

Semi-empirical correlations [11] can calculate the wrinkling factor K. In Quick-Sim the wrinkling factor is calculated locally using a phenomenological approach (Weller-model). On the other side, the laminar flame speed strictly depends on the chemical interaction between fuel and air, but as well from the thermodynamic engine conditions before and during the combustion event. For this reason, the laminar flame speed is calculated through detailed chemical reaction schemes and tabulated into a look-up table. The engine flow field characteristics (pressure, temperature, etc..) calculated during the simulation are used to read the corresponding value of the laminar flame speed into the look-up table. Therefore in QuickSim, the laminar flame speed is calculated as a function of the mixture composition, pressure, temperature (T_U), EGR fraction ($w_{EGR,U}$) and water concentration ($w_{Water,U}$) of the unburned gas as shown in eq. 3.3.

$$S_L = f(\lambda, p, T_U, w_{EGR,U}, w_{Water,U})$$
eq. 3.3

In all the engine simulations realized in the current work, chemical reaction mechanisms replace the semi-empirical formulation from Gülder [34] to calculate the laminar flame speed, which allowed to consider the influence of the fuel composition in the simulations. Even though a traditional RON95 fuel was adopted for the single-cylinder engine investigation, the Knock analysis requires an accurate fuel model in the 3D-CFD simulations. The kKock model and its calibration are discussed in detail in chapter 6.

3.7 Spray modelling

The quality of a spray model is crucial for the correct reproduction of the engine processes through 3D-CFD simulations, as mentioned in chapter 2. Besides, the selection of a RANS approach as CFD methodology in QuickSim requires the choice of a consistent injector model that could be integrated in the full engine simulations (usually CFD injector models are more detailed and based on LES approach as it can be found in literature [46]).

Chapter 4 describes in detail the integration of a simplified injector model, based on a RANS approach, a mixed Euler/Lagrange formulation and the neglect of the primary break-up, while maintaining an acceptable level of precision. Furthermore, chapter 4 provides the description of a new methodology for the spray targeting calibration within the full engine development.

4 Simulation of Injections

Beyond the implementation of water physics in QuickSim, further development of the injection model was accomplished. The improvements of the injection model were realized by coupling optical measurements performed in a constant volume chamber and by the setting of a new methodology to calibrate the simulations with a repeatable analysis of both experimental and numerical results. Axial and radial penetration curves, injection tip velocity, as well as jet angles, were measured and evaluated. The injection model calibration realized through these data, enhance the simulations' reliability. Finally the modelled injectors were tested in the single-cylinder engine simulation described in chapters 5, 6, 7 and 8.

4.1 Problematics of spray analysis

Injection modelling is one of the most critical points when simulating internal combustion engines, due to the high number of influencing factors and the complex statistic nature of the injection event itself. The goal of the injection modelling in the 3D-CFD tool QuickSim bonds the need for a good description of the spray together with a time-effective integration with the simulation of several engine cycles. As results, a successful integration and calibration of the injection models call for a new methodology which considers input parameters coming from optical measurements. The methodology mentioned above was developed at FKFS Stuttgart, together with Mr S. Hummel and Mr E. Rossi. For a detailed description, the reader is addressed to [50].

The literature about injection analysis assesses the importance of a good agreement between fuel modelling and spray reproduction for the numerical characterisation of the injection phenomenon [57]. In addition, fuel evaporation is a crucial event for the correct simulation of engine behaviour and it depends as well on the interaction between the injected fluid and the engine flow field.

Therefore, it is necessary to simplify the injection model, to take into account these aspects, but neglecting the internal injector dynamics and initialising the droplets inside a region of the mesh. To correctly initialise the injection, the droplets and velocity distribution are adjusted to imitate the spray behaviour observed through the measurements [53].

However, the volatility of the gasoline in a constant pressure chamber does not reflect real engine conditions (use of n-heptane, lack of motion field and constant pressure and temperature conditions are usually the main differences compared to injecting during the engine compression stroke) [53]. For these reasons the data provided by a constant volume chamber can be used to understand the behaviour of fuels and of injectors and to define the relation between the more important injection parameters such as, penetration depth, injector targeting, jets interaction, etc.. These data must therefore be read critically and not considered in absolute terms. Finally, the relationships found between the different parameters influencing the injection can be used to calibrate the simulations.

The validation of the goodness of an injection model also passes from the final behaviour of the model when it is considered in a real engine flow field. In fact, a model that achieves satisfactory results in a constant pressure chamber does not mean that it will produce reliable results even in engine simulations, especially since the break-up and vaporization model are used under different conditions. For this reason, after validation of the model in a constant pressure chamber, it is necessary to analyse the results of the model in a real field of motion.

Unfortunately, it is extremely complicated to carry out optical measurements of an engine at the level of mixture analysis and then pair them directly with engine simulations. However, testing the injection model in an engine simulation has the advantage of being able to use engine parameters that are easy to measure on the test bench to understand in general the goodness of the model when applied under real conditions. In fact, the tendency of the droplets to break them up, mix and vaporize can correlate engine parameters such as lambda, combustion speed, pollutant formation and specific power, which are directly related to the mixture generation and therefore to the injection event. Therefore, for a global simulation approach, the spray model must be first validated through optical measurements in a constant pressure chamber and then it has to be stressed within the engine flow field, considering macro-engine parameters.

The starting point is the spray characterisation in the CFD simulations through the attempt to reproduce optical measurements with direct image comparison. However, this methodology is sometimes rough and barely reproducible. Starting from the fuel and the injector modelling presented in previous works [11, 56], the injector simulations were improved through the analysis of optical measurements, by means of an imaging tool, to detect the main spray features and by providing these parameters as inputs for the simulations [50]. This chapter investigates several parameters influencing the reproduction of injection event, such as tip speed, velocity distribution, droplets initialisation range, to enhance simulation predictability.

4.2 Injector characteristics and optical measurements

The optical measurements used in the following analysis were carried out by Mrs M. Gern at TU Berlin during the FVV-Project n.1256. The high-speed video measurements, based on shadow imaging, were performed for the multi-hole injectors described in table 4.1. These correspond to the injectors investigated at the single-cylinder engine test bench. *Injectors 1* and *Injectors 2* were used for water injection operations, respectively for IWI and DWI, while *Injectors 3* was adopted for direct gasoline injection. Nevertheless, the possibility to move the direct lateral water injector (*Injectors 2* in table 4.1), corresponding to test bench configuration, to the central position (for DWI) will be analysed in chapter 8, through the virtual test bench.

Table 4.1: High pressure and low pressure injector features

Injector	Position in cylinder head	Inj. Pressure [bar]	Holes [#]	Fluid
1	Intake runner	≤ 10	8	Water
2	Direct, lateral	≤ 200	5	Water/Gasoline
3	Direct, central	≤ 200	5	Gasoline

The description of the injection model calibration methodology is only reported for *Injectors 2* since it was the one tested for both gasoline and water injection. The calibration methodology refers to the compare between optical measurements and corresponding simulation in the constant volume chamber.

The novelty of the analysis is the proposal of an integrated approach for testing different injectors virtually in a real engine flow field. In order to make this methodology working, it was necessary to support experiments conducted in a constant volume chamber with the respective simulations and finally to extend these results to full engine simulations. Moreover, this methodology allowed a rapid validation of the models by coupling optical measurements to CFD simulations.

4.3 Models used and parameters description

The jet atomization is a highly complex process due to the numerous influencing factors such as the nozzle geometry, the fuel specification, discharge velocity and the interaction with the background fluid, pressure and temperature conditions [44]. The primary droplet breakup consists of the formation of ligaments and droplets from a coherent liquid phase, close to the injector orifice. It mainly depends on jet velocity and nozzle flow. The injection model in QuickSim does not include the primary droplet breakup and the internal injector flow [11, 50, 55]. To overcome this simplification literature recommends to thoroughly calibrate the secondary droplet breakup parameters as well as the initial droplet size through optical spray measurements [3]. The employment of this approach has some limits, such as the simulation of supercritical injection conditions, which result in a supersonic gas jet. Such a simulation would require a highly resolved computational mesh, having a cell discretization length that is even smaller than the cross-sectional width of the injector orifice [6].

It is, therefore, necessary to select appropriate initialisation parameters still to capture the fuel propagation within the combustion chamber. For the purpose of virtual engine development, the injection simulation is integrated into the engine cycle simulation, meaning that it has to be combined with a RANS calculation method, being anyhow able to ensure a certain accuracy in reproducing the real phenomena. For this reason, the injector model was designed to ensure an easy coupling between model inputs and optical measurements to increase the accuracy of the model, while maintaining acceptable calculation times. Considering figure 4.1, the computational mesh of the constant volume chamber is described.

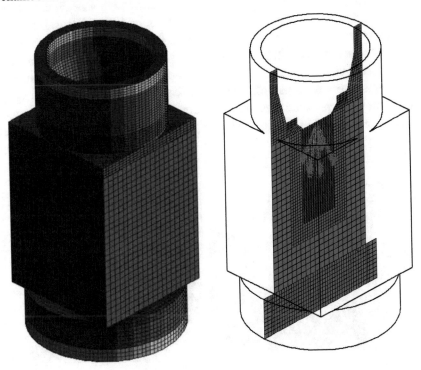

Figure 4.1: Mesh of the constant volume chamber built in QuickSim on the left side and spray section with mesh refinement on the right one.

As said before, QuickSim does not require a high-refined grid but it rather employed structured hexahedral coarse meshes. Only 113000 cells are used to reproduce the pressure chamber for the region where the fluid is injected, with a discretisation length of 1.5 mm (the chamber volume is 3.5 litre). Boundary conditions are applied far from the analysed volume of interest.

When considering the low-pressure water injector, the experiments were carried out using a different test rig, with respect to the high-pressure chamber shown before (see figure 4.1), realized as well at TU Berlin. This device consists of a tube at the end of which an injector is mounted, while on the opposite side a suction system drags the injected fluid away with a negligible speed compared to the speed of the injected droplets. As a result, a new simulation environment was built in order to reproduce the new measurements system. The right side of figure 4.2 shows the new mesh in comparison with the real test rig (left side of Figure 4.2). As inlet boundary condition, pressure was applied and to reduce the influence of the boundary conditions on the spray simulation, the pressure was imposed through enlarging the mesh of the real tube as reproduced in the right side of figure 4.2. On the other hand, as outlet boundary condition, a mass flow was used in order to replicate the flow field of the real test rig.

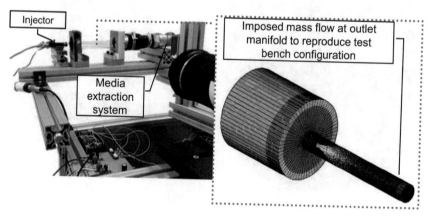

Figure 4.2: Reproduction of the low pressure water injector test rig installed at TU Berlin in the CFD environment of QuickSim.

4.3.1 Spray definition and geometrical parameters

The mesh does not integrate the injector itself physically, except for its housing, but just a coordinate system defines the injector position in QuickSim. The integration of the injector geometry and the reproduction of its internal flow can better reproduce local processes, but this hits against the problem of a time-effective integration of the injection phenomena within a full engine simulation and willingness to provide internal geometry data from injector manufacturers. The injector spray geometry is modelled based on four characteristics angles, which define the direction of the spray jets, for multi-hole injectors, or the jet width, for hollow-cone injectors. These angles are shown in figure 4.3 [11, 55]. The zenith angle θ identifies the radial distance from the injector middle axis, and the azimuth angle ϕ describes the jet orientation starting from the reference direction, given by x-axis. The angle α adjusts the width of the jets. For hollow-cone injectors the jet middle axis coincides with the injector middle z-axis; thus zenith and azimuth angles are equal to zero. One additional angle γ is used to reproduce hollow-cone sprays, while the angle α corresponds to half of the spray cone angle.

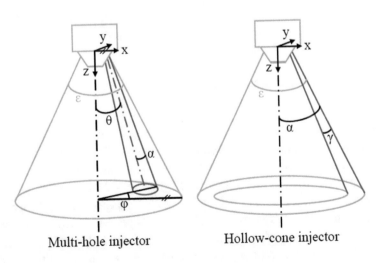

Figure 4.3: Geometric spray definition in QuickSim [11] [55].

The combination of the geometrical definition of the initialization region for the location of the initial droplets, together with the SMD and the exit velocity allow defining the desired spray geometry in accordance to the given spray targeting resulting from optical measurements. Since the neglection of internal injector flow and the primary breakup, the droplets physical properties are initialized in a conical region which is defined through the parameters Lmin, Lmax and a characteristic angle, as shown in figure 4.4 [11] [55].

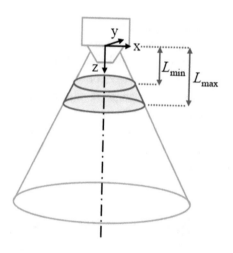

Figure 4.4: Spray initialization domain [11] [55].

A well-defined number of non-interacting droplets, with identical physical properties, is grouped into samples, called parcels. The calculations consider for engine simulations a total of 130000 parcels, meaning that for example by injecting 80 mg fuel, an initial SMD of 12 μm (by 200 bar injection pressure), a conventional gasoline ($\rho = 670 \ kg/m^3$ at 25°C temperature) and supposing spherical droplets, approximately 1000 droplets per parcel are taken into account. The influences of this parameter were investigated in [55]. The author demonstrated that a good compromise in terms of computational time and reliable reproduction of penetration and evaporation behaviour achieves with a thousand droplets per parcel. A high number of droplets and therefore a small number of parcels thins out the spray significantly and results in an underestimation of the spray penetration [55]. A decrease in the number of the droplets, on the contrary, generates a highly crowded spray [55].

The calculation framework in QuickSim consists of a blend formulation between Lagrangian and Eulerian approach, where the conservation equations of mass, momentum and energy for the dispersed phase are written for each parcel. The Lagrangian approach was selected because it enables to reproduce high velocities in the nozzle vicinity without significantly influencing the physical processes [57]. A Rosin-Rammler function is used to initialize the diameter of each droplet, as a function of the measured SMD. The definition of the individual droplet properties represents the calibration endeavour which influences on the secondary break-up, and the phase exchange process for mass, momentum, and energy, according to the standard Euler-Lagrange formulation of multi-phase flow. Throughout the whole injection event, the software varies all the initialization input quantities statistically independently and continuously for each time step. The variation of these parameters, together with the experimental data from a constant pressure chamber, tune the penetration curve in the simulations.

4.3.2 Breakup model

Once the droplets are generated into the computational mesh (by neglecting the primary breakup) the interaction with the surrounding media (secondary breakup) is of fundamental importance to reproduce the jets progression, the air entrainment and thus the mixture formation. Since the goal is to connect the characteristics of the jet with the quality of the mixture formation, the evaporation process was studied considering the fuel properties and the velocity between the droplets and the surrounding gas. For these reasons, different breakup models were screened, using the agreement between the characteristic angles of the jet and its penetration as an evaluation parameter among experimental data and simulations [45].

The breakup of a liquid jet depends at first on the discharge velocity and its influence on the (aerodynamic) forces affecting the jet. Once the liquid jet exits the nozzle, its stability is related to the surface tension, the internal forces of the jet and the destabilizing forces induced by the interaction with the surrounding medium [42, 51]. As usually done to reproduce phenomena which depend from the interaction of different parameters, it is useful to describe the droplet breakup by means of dimensionless quantities [16, 42]. Therefore three characteristic numbers are usually used to described breakup regimes:

- **Weber number** *We*

  It relates inertia forces of a fuel droplet with its surface tension σ_f.

$$\text{We} = \frac{\rho_g \cdot d_D \cdot v_{rel}^2}{\sigma_f} \qquad \text{eq. 4.1}$$

ρ_g	Density of the surrounding gaseous fluid
d_D	Droplet diameter
v_{rel}	Droplet relative velocity

- **Reynolds number** *Re*

  It considers the effect of the dynamic viscosity η_f of the liquid serving as stabilizing factor against the breakup. The Reynolds number (already mentioned in chapter 2 and conveniently here reported with a coherent formulation for the droplet problem) indicates the ratio of inertia and viscous forces.

$$\text{Re} = \frac{\rho_f \cdot d_D \cdot v_{rel}}{\eta_f} \qquad \text{eq. 4.2}$$

- **Ohnesorge number** *Oh*

  It resumes the relation of the viscosity to the surface tension of the fluid and is strictly dependent on the fuel characteristics.

$$\text{Oh} = \frac{\eta_f}{\sqrt{\sigma_f \cdot \rho_f \cdot d_D}} = \frac{\sqrt{\text{We}}}{\text{Re}} \qquad \text{eq. 4.3}$$

The so-called *Ohnesorge diagram* reported in figure 4.5 shows the different spray breakup regimes, including the atomization regime, as a function of the Reynolds and Ohnesorge number. In [42] Pfeifer assesses that by constant Ohnesorge number, i.e. consistent fuel properties and injector geometry, an increase of the discharge velocity (due to higher injection pressure) and therefore of the Reynolds number, leads to a higher order of spray breakup.

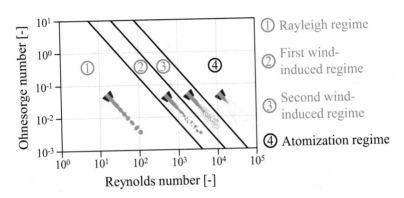

Figure 4.5: Classification of the spray breakup regimes through the diagram of Ohnesorge[42].

In the current analysis, the adopted breakup model is that of *Reitz and Diwakar*. According to this model, droplet breakup occurs due to aerodynamic forces acting at the interface between liquid droplets and the surrounding air. The action of these forces can be modelled as wave perturbations on the droplets' surface, characterized by a specific growth rate and wavelength. The instability generated by these perturbations causes the droplet breakup, thus reducing the droplet's diameter. The rate of the diameter reduction depends on the combination of surface tension and inertia forces [44]. The model considers two possible breakup regimes [10]:

- **Bag breakup**: non-uniform pressure field around the droplet causes it to expand in the low-pressure wake region and eventually disintegrate when surface tension forces are overcome. Instability is determined by a critical value of Weber number.

$$\text{We} = \frac{\rho |u - u_d|^2 d_D}{2\sigma_D} \geq C_{b1} \qquad \text{eq. 4.4}$$

where C_{b1} is an empirical coefficient usually set $C_{b1} = 6$. The droplet size which satisfy equation eq. 4.4 is the stable droplet size and the corresponding characteristic time is reported in eq. 4.5, where $C_{b2} \approx \pi$.

$$\tau = \frac{C_{b2}\rho_d^{\frac{1}{2}} d_D^{\frac{3}{2}}}{4\sigma_D^{\frac{1}{2}}} \qquad \text{eq. 4.5}$$

- **Stripping breakup**: process in which liquid is sheared or stripped from the droplet surface. The condition for the onset of instability is given by eq. 4.6.

$$\frac{We}{\sqrt{Re_D}} \geq C_{s1} \qquad \text{eq. 4.6}$$

where Re_D is the droplet Reynolds number and C_{s1} is a coefficient with default value of 0.5. The characteristic time scale is reported in equation eq. 4.7.

$$\tau = \frac{C_{s2}}{2} \frac{\rho_d^2}{\rho} \frac{d_D}{|u - u_D|} \qquad \text{eq. 4.7}$$

where the adjustable coefficient C_{s2} can assume value between 2 and 20.

These instability mechanisms determine the corresponding droplet breakup rate with respect to a characteristic breakup time τ. Notably, the rate of reduction of the droplet diameter is calculated through eq. 4.8.

$$\frac{d\,d_D}{dt} = -\frac{(d_D - d_{D,stable})}{\tau} \qquad \text{eq. 4.8}$$

Just like the characteristic time scale, also the stable droplet size depends on the breakup mechanism. It can be determined through the flow conditions considered in the Reynolds and Weber number, as well as the fuel specifics, captured in the Ohnesorge number [16].

4.3.3 Droplet evaporation

During the droplet evaporation, which is competing with the secondary breakup mechanisms, the liquid fuel continuously changes into its gaseous state, forced by ambient temperature and pressure conditions. Each pressure is assigned a specific temperature at which the fuel evaporates, and specifically, these are the saturation or vapour pressure p_s and boiling temperature T_b respectively.

Changing pressure conditions result in a variation of the boiling temperature. For pure substances, this relation is well characterised by the saturation pressure curve. However, liquid fuel like gasoline consists of hundreds of components with varying boiling temperatures, mainly depending on their number of carbon atoms. The molecular interaction of these components delineates the fuel-specific boiling curve, which describes the evaporation rate according to the temperature at a constant pressure level [16, 52].

The process of evaporation is strongly influenced by the fuel properties and particularly encouraged by high temperatures, a high partial pressure gradient as well as high relative velocities between the droplet and the surrounding gas (enhancing the convective mass and heat exchange [16]). The relative velocity, in turn, is supported by injection pressure and by engine flow field characteristics (e.g. high Swirl or Tumble motion). In opposition, break up mechanisms cause smaller droplet sizes which brings an increase of the specific fuel surface and therefore its evaporation rate.

Exchange processes between the droplet and the surrounding gas occur based on conservation laws, represented by the equations for mass, momentum and energy. During the evaporation, mass transfer, which causes a decrease of the droplet size, is accompanied by heat transfer due to the differences in temperature. The thermal energy is extracted from the surrounding gas, resulting in a reduction of temperature (if the droplet's temperature is lower than the one of the gas). This effect is even more evident for water due to its six times higher evaporation enthalpy with respect to gasoline, as described in chapter 1.2.2.

Besides, QuickSim includes an evaporation model that takes into account the effect of *Flash Boiling*. This event is a sudden droplet burst and immediate evaporation caused by an abrupt decrease of the ambient pressure below the temperature-dependent saturation pressure of the fuel component. Generally, it occurs when $T_{droplet} > T_{critical}$ such as , for example, if heated fuel exits the injector into a low-pressure regime. The *Flash Boiling* can have a significant influence on the characteristic spray pattern and propagation.

Depending on the injection timing, different influencing factors on the fuel evaporation are prevailing. High relative velocities characterise direct injection of fuel during the intake stroke due to the strong charge motion (high Tumble level, turbulence). During the compression stroke, however, the in-cylinder pressure and temperature level increases, whereby the temperature rise represents the dominant influencing factor [16].

With resuming purposes, secondary breakup mechanisms cause a reduction in the dimensions of the droplets which lead to an increase in the specific fuel surface and in the rate of the evaporation process of the liquid which continuously changes into a gaseous state according to the ambient temperature and pressure conditions. The fuel properties strongly influence the evaporation process and the gradient of pressure and velocity between the droplets and the surrounding gas. These aspects were programmed to be part of the injection model.

One limit of the breakup model settings of the software is to neglect inter-droplet collisions and coalescence. Even though these effects are important for reproducing the real spray micro behaviour, they have minor influences on the overall engine behaviour and mainly, once the spray is analysed within the flow field of the combustion chamber. On the other hand, these interaction effect between the droplets are quite important when trying to predicting emission formation. Therefore, these submodels are currently under investigation considering their effect on the spray development and their cost in terms of computational effort.

The experiments needed to calibrate the model are reduced if the manufacturer provides the spray targeting and the static mass flow of the injector. Generally, two injection pressures in combination with two chamber back pressures are recommended to predict the jet speed at the injection tip correctly. In addition, a temperature variation of the injected fluid and of the chamber is suggested. The calibration of the tip speed is the most influencing parameter for reproducing the jet penetration and its evaporation. This parameter is the most critical quantity to be individuated, and its accurate prediction is also the main limitation of the model since the neglection of the internal injector flow field [50].

The objective of the injection model is not only to emulate the spray characteristics under stationary conditions but also the correct simulation of the injection event within the flow field generated by the engine compression stroke. Therefore the accuracy of the model is again tested within the full engine simulations, verifying the ability to predict the main engine parameters such as lambda, heat release rate and burn rate correctly in comparison to test bench results.

4.4 Imaging tool and injector calibration methodology

Alongside 3D-CFD simulations, an imaging software was built and validated through different experimental test campaigns led at TU Berlin and at IFS [45], which allows processing *Mie-scattering* and *Shadow imaging*.

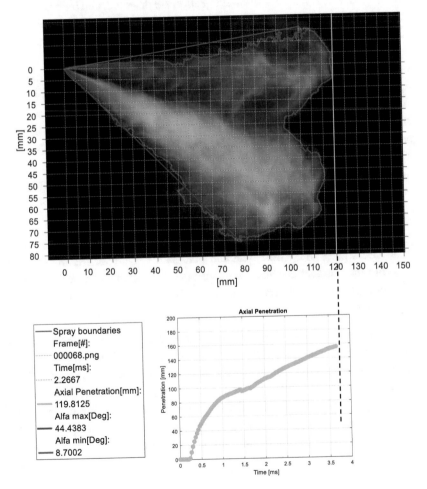

Figure 4.6: Imaging tool applied to Shadowgraphy for the calculation of the characteristic angles and penetration of the spray.

The purpose was to create a repeatable methodology to analyse spray formation and propagation using the same tool to evaluate measurements and 3D-
CFD simulations. Two different scripts run parallel over the measured and the
simulated injection respectively and both evaluate the level of luminance for
the injection frames. The injection electric signal is the camera trigger to start
recording. The early frames without injection are detected, and the background
is stored to be subtracted to the frame where the injection happens. This mechanism is created to delete any source of light except for the reflecting fluid jet
and to evaluate the injector's hydraulic delay with respect to the electric signal.
A scaling factor is automatically applied by recording a characteristic length
in the injection chamber. The videos are transformed into greyscale images
and then converted to binary using a variable thresholding algorithm, based on
Otsu [38]. The binary frames are collected into one matrix, and for each of
these, the boundary pixels are stored tracing the exterior jet contour. Each row
of these matrices contains the row and column coordinates of a boundary pixel.
The tool then automatically characterises the jet for each injection timing, computing axial penetration, tip speed and angles corresponding to maximal upper
and lower vertical penetrations, as represented in figure 4.6.

Figure 4.6 establishes a time and spatial resolved analysis of the injection event,
which investigates the geometrical and kinematics characteristic of the spray.
A recent improvement elevates the reliability of the analysis by performing an
imaging average of 5 different shootings, before calculating the main geometrical parameters.

Among the boundary regions detected, through numerical methods, the algorithm can individuate the primary spray contour or the different jets composing the spray, depending on the injector used. Once the experimental data have
been collected and characterized, the calibration of the 3D-CFD simulation is
realized by applying the same tool with a customized script on the simulation
outputs also evaluating the main geometric parameters of the spray. About
figure 4.7, the experimental results and the simulations considering the above
mentioned geometric parameters are compared, and recursively the numerical
analysis can be improved.

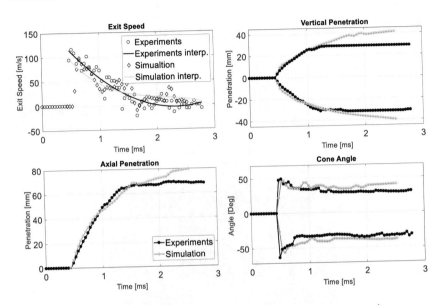

Figure 4.7: Calibration and validation of the simulation by comparing geometrical parameters and optical measurements.

Beyond the imaging analysis, an index of the coherence of the jet can be calculated, as a first assumption, by monitoring the trend of the maximum upper and lower penetration angles. The detected angles are treated statistically, bearing in mind a coefficient of variation that can be defined as the ratio between the maximum angle standard deviation and its mean value along with the injection. The smaller the coefficient of variation, the less dispersed the data and thus the higher the coherence of the jet. This methodology was applied for the three different injectors described in table 4.1. Using this approach, the respective injector models were calibrated in the 3D-CFD simulation environment, varying injection pressure (from 100 to 200 bar) and chamber back pressure (1 to 5 bar). Considering this automated procedure, the effort for the simulation calibration was reduced, and the model predictability enhanced.

4.5 Different behaviour between water and gasoline injection

The injection of water was studied with respect to those of gasoline to invest-igate the different behaviour of the two fluids when using the same injection device. With reference to figure 4.8 gasoline and water injection were com-pared at 150 bar injection pressure and different chamber back pressures [20].

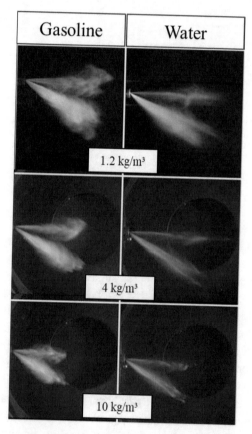

Figure 4.8: Results of the DWI optical spray measurements using water and heptane at 150 bar injection pressure and for different chamber densities [20].

The temperature of the injected fluid was conditioned at 40 °C. The imaging tool described in chapter 4.4 analyses the high-speed videos and the outcomes of the study were used, on one side, to calibrate the spray model and to validate it also for water injection application, and on the other side, to quantify the real impact of the fluid choice on the injection strategy.

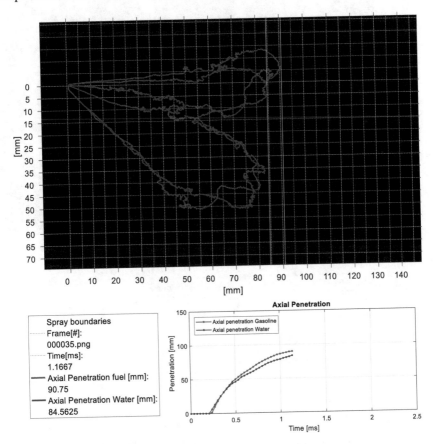

Figure 4.9: Imaging tool overlapping water and gasoline injection, using *Injectors 2* (see table 4.1) at 150 bar injection pressure and considering 1.17 ms from electrical SOI.

For the high-speed measurements, distilled water was considered against gasoline (commercial RON98) and the videos were evaluated according to the recommendation of the SAE standard [27].

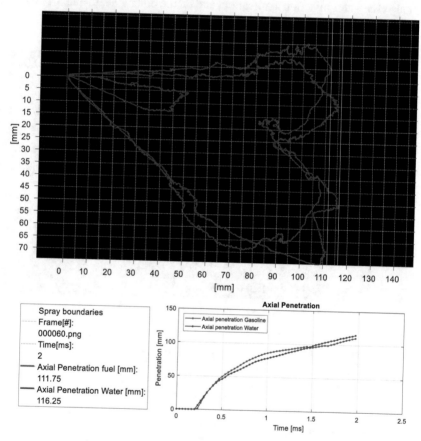

Figure 4.10: Imaging tool overlapping water and gasoline injection, using *Injectors 2* (see table 4.1) at 150 bar injection pressure and considering 2 ms from electrical SOI

Figure 4.9 represents an example of analysis led on injection measurements through the self-developed imaging post processing tool. It refers to the case of 150 bar injection pressure, ambient temperature and pressure condition in the chamber and 1.17 ms from electrical start of injection (SOI). The high-speed videos were reduced into frames via the imaging tool, identified the limits of the jet at each time step, separately for water and gasoline injection. Then the real jet image was subtracted, and the boundary generated by water were overlapped with the ones of gasoline, producing the analysis shown in figure 4.9.

Through this methodology, it was possible to quantify the hydraulic delay of the injector using the two fluids. By injecting water, the hydraulic delay was detected as almost comparable with the one of gasoline, despite the different fluid density and viscosity. The electric signal was set to 5 ms and in 0.25 ms the fluid exited from the nozzle (the delay could be quantified as 5% to the duration of the electric signal).

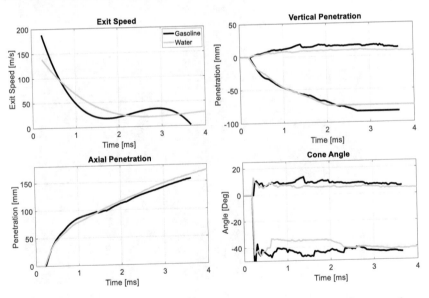

Figure 4.11: Comparison of exit speed, penetration and angles for water injection and gasoline injection at 150 bar injection pressure and using *Injectors 2* (see table 4.1)

After 0.5 ms gasoline penetrated more than water, which was an unpredicted behaviour. Water has higher surface tension and it is expected to have higher penetration. Nevertheless, the higher density of water could initially slow down the first exiting of the jet, and it generated an utterly different injector internal dynamic.

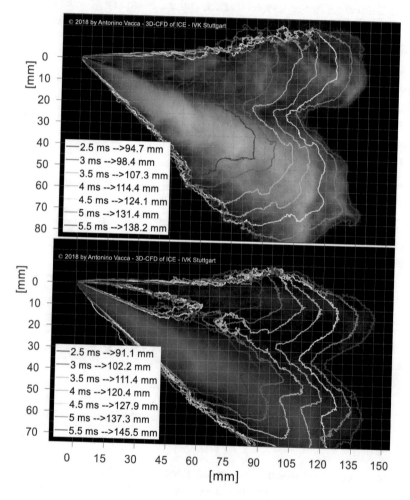

Figure 4.12: Time and spatial resolved analysis of the injection event for gasoline (up) and water (bottom) at 150 bar injection pressure and ambient chamber conditions.

After 1.5 ms from the start of injection (electric start of injection), the higher surface tension and the more significant droplet dimension let finally water realizing a global higher penetration. Concerning figure 4.10, the comparison between the two fluids is reported, at 2 ms injection time. The water jet is more compact, it has lower global angles, and it finally has higher penetration than the gasoline one. In order to quantify the different behaviour between water and gasoline, the imaging tool provided a comparison in terms of exit speed, vertical penetration, cone angles and axial penetration as reported in 4.11. These trends were given as inputs in the CFD simulation as injector initialization parameters, enhancing the reliability of the simulations.

Figure 4.12 finally provides a comparison of the jet development for gasoline and water at 150 bar injection pressure, using the later direct injector again (see table 4.1) This represents a time and spatial resolved analysis of the all injection event. Each colour corresponds to the boundary limits at the most meaningful time step and the respective penetration. From the spray contour development, it can be seen that water has higher penetration than gasoline, starting from 2.5 ms. As mentioned before, the water jets are more compact to gasoline, producing a completely different plume. The penetration curve, angle development and jet shape were directly transferred as input in the injection model and used for the full engine simulations.

4.6 Spray targeting developments and engine virtual testing

The purpose of the measurements carried out with the injection chamber was to characterize the different injectors and to provide the data for the calibration of the injection model into the 3D-CFD software QuickSim. On the other side, as said, this injection model aims to be integrated into a full engine simulation. Specifically, the robustness of the model has to be tested in a real engine flow field, where the gas exchange processes through the valves, the piston stroke, as well as the combustion event, are considered. For this reason, the injector development methodology was tested in the full engine model described in chapters 5, 6 and 8, which deal with the application of water injection to modify the base engine map on three operating points. As a result, a full description of the single-cylinder engine model will be given in chapters 5, 6 and 8, while in the current section the focus is still the investigation of the injector development approach and of the influence of the injection model on the global engine parameters However, since the model must be universally applicable, particularly both for injection and engine simulations, a comparison with the engine simulations is performed for the calibration of the injector model itself.

The engine model mentioned above corresponds to a single-cylinder research engine investigated at the test bench for water injection experiments [48]. It was equipped with direct fuel injection, through a conventional solenoid valve for high-pressure fuel injection, located centrally in the cylinder head. The engine data and configuration will be resumed conveniently in chapter 5. In the current section, to test the injection model, the operating point 2000 rpm and 20 bar IMEP (OP1) was selected. OP1 corresponds to maximum torque at this engine speed. The engine was regulated at Knock limited conditions with a spark advance of 1° CA b.FTDC. A delayed spark ignition characterizes this load point due to the onset of Knock. In order to enhance the Knock resistance and shift the combustion towards early phasing, one possible solution is to improve the mixture formation and to exploit the fuel evaporation and distribution as much as possible to reduce the in-cylinder temperature. For this reason, an embedded approach for the development of the spray targeting within the real engine flow field is needed.

The so-called *Lambda map* helps to study the mixture formation. With reference to figure 4.13, it is possible to analyse the air to fuel ratio of the engine at each crank angle, during the compression stroke, bearing in mind the real 3D engine flow field and calculating, for example, the differences produced by various injection strategies. Specifically, at each crank angle, the volume of the combustion chamber is calculated, and the air to fuel ratio is grouped into different bins within the current volume. For each crank angle then the percentage of the cylinder volume at a specific lambda can be identified. From the 3D map and the left y-axis, it can be gathered that with *Injector 3* less than 1 % of the volume at firing top dead centre (FTDC) presents exactly a stoichiometric air to fuel ratio ($\lambda = 1.000$).

It is then useful to group the percentage of volume at each lambda into bands of multiple lambda values. Considering again figure 4.13 and the right y-axis, the dotted lines group the percentage of volume at a particular lambda into three bands:

- rich volume percentage (red line)
- stoichiometric volume percentage (blue line)
- lean volume percentage (black line)

The stoichiometric band corresponds to the mixture characteristic pursuits to increase the combustion speed. With the base injector (*Injector 3*) the engine features 15 % of its combustion chamber volume in optimal conditions (λ : $0.85 \Rightarrow 1.05$) at FTDC, while 30 % of the volume has rich mixture and 55 % is in lean condition.

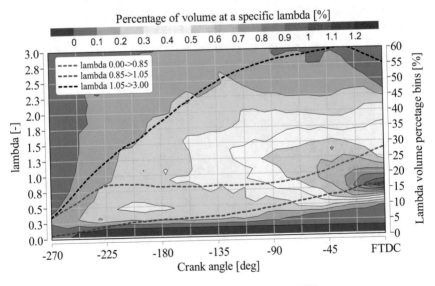

Figure 4.13: *Lambda map* for fuel *Injector 3* at OP1 [50].

However, the mixture formation and the time history of the homogenization event must not be considered separately concerning some other essential aspects that could prevent the injector from being successfully used in nowadays engines. Among this, a crucial role in the evaluation of the spray characteristics is played by its tendency of generating wall impingement issues and consequently, the rise of emission formation and oil dilution. In order to judge the overall behaviour of the investigated injector targeting, it is worth to quantify in this sense, the amount of liquid fuel impinging the cylinder liner. For this purpose, a post-processing script allows isolating the CFD-cells of the liner from the rest of the mesh and calculating the liquid fuel mass passing through this region [45, 50]. A good compromise between a wider spray and wall impingement issue represents the challenge for the design of an effective spray targeting. For this purpose, the support of 3D-CFD simulation offers an essential and in this case, time-effective support, especially coupling the spray analysis with the engine response.

The application of this methodology will be used to study as well the behaviour of water injectors within the engine simulations (chapters 6 and 8). Among other parameters, the water distribution and its tendency to impinge the cylinder wall will be considered as essential aspects for the choice of the most effective water injection strategy.

5 Optimization of the Engine Map through Water Injection

The current chapter provides a description of the experimental and simulation methodology, especially with a focus on the investigation of the thermodynamics of a single-cylinder engine whose date will be resumed in the next section (chapter 5.1).

More in detail, three engine operating points will be optimized through the adoption of different water injection strategies. Table 5.1 resumes the load points considered and the respective optimization targets.

Table 5.1: Investigated engine operating points for the optimization of the single cylinder engine at the test bench and the respective targets.

Operating point	Speed [rpm]	IMEP [bar]	λ [-]	Target
1	2000	20	1	Knock reduction/increase of compression ratio (ε)
2	2500	15	1	Emission control
3	4500	20	0.85	Prevent fuel enrichment

The base calibration of *Operating point 1* (OP1) was characterised by a delayed spark timing, 1°CA b.FTDC , in order to prevent Knock occurrences. The engine was always regulated with a stoichiometric mixture. Therefore the main objective of the investigation for OP1 was the possibility of using water injection for enhancing efficiency by generating higher Knock resistance. The solutions analysed to achieve this purpose were indirect and direct water injection, Miller cycle and a combination of them. On top of that, through the Knock analysis, it was possible to evaluate the increase of the engine geometrical compression ratio virtually.

The *Operating point 2* (OP2) base calibration already consisted in an optimized ignition point (10°CA b.FTDC) and the engine running at stoichiometric conditions. Some attempts to increase the indicated efficiency of this load point were done with IWI and DWI strategies, but no solutions were found to further reduce the indicated specific fuel consumption (ISFC). On the other side, there was the possibility to reduce the engine row emissions and to investigate the pollutant formation with respect to the water injection strategies. These results are reported in detail by Mrs M. Gern in [21] and just the influence on Soot formation to the water injection timing are described in the current work.

The *Operating point 3* (OP3) represents the engine rated power. At this load and speed the engine is requested to control the maximum T_3 through the injection of additional fuel, thus a fuel enrichment corresponding to $\lambda \simeq 0.85$ was needed. The single-cylinder engine at the test bench was affected by cylinder head overheating during all measurements campaign. Thus to prevent any engine damages, it was decided just to investigate OP3 through 3D-CFD simulations and particularly with the support of the *Virtual Test Bench*, described in chapter 5.2.

The base engine calibration for the above-mentioned load points is resumed in table 5.2. These parameters were kept unchanged unless specified in each chapter. Also, the engine back pressure was always set at 1 bar. The cam phaser was kept to rest position corresponding to 51 °CA valve overlap, and 194 °CA intake/exhaust valve event length at 1 mm lift.

Table 5.2: Base engine calibration for the three investigated operating points.

OP	Valve overlap [°CA]	Injection pressure [bar]	SOI [°CA b.FTDC]	$p_{2.2}$ [bar]	IP [°CA b.FTDC]
1	51	150	300	1.65	1
2	51	150	300	1.3	9
3	51	200	320	2	4

5.1 Set-up of the single-cylinder engine test bench

The single-cylinder engine at the test bench derives from a 3-cylinder gasoline direct injection engine currently in the market. This engine was installed at TU Berlin and characterised in the project n.1256 supported by the Forschungsvereinigung Verbrennungskraftmaschinen (FVV), in order to investigate indirect and direct water injection operations. The main engine features are resumed in table 5.3.

Table 5.3: Technical data of the single cylinder research engine.

Displaced volume	333cc
Stroke	82 mm
Bore	71.9 mm
Compression ratio	10.75:1
IWI pressure and timing	$\leq$ 10 bar, 360°CA b.FTDC
DWI pressure and timing	$\leq$ 200 bar, 120°CA b.FTDC

It was equipped with optical access and an external boosting system which could be independently regulated, together with a back pressure valve simulating the presence of turbo compound. The intake manifold was specially manufactured to include a single water injector 80 mm away from the intake valve, corresponding to *Injector 1* described in chapter 4.1. The cylinder head was built to house an additional lateral injector, used just with water (*Injector 2* reported in chapter 4.1). In contrast, the gasoline injector was placed in a central position, as it is in the production engine configuration (*Injector 3* as analysed in chapter 4.1). The manifold injector was tested up to 10 bar injection pressure (15 bar injection pressure tested just through simulation) with different injection timing, while direct injection operations were led up to 200 bar injection pressure. In the current work, emulsion solutions were not considered.

The low-pressure injector (*Injector 1*) was an already water-developed device, while the lateral direct water injector (*Injector 2*) was a conventional multiholes gasoline device. For this lateral injection position, an injector with an asymmetrical spray pattern was selected.

The Knock event was experimentally determined when the Knock ratio (KRAT) overcomes the value 1.1, considering 100 working cycles. This index represents the ratio between the level of the high pass filtered pressure signal and the low pass filtered reference level, before the maximum pressure peak is reached, therefore characterising the Knock intensity.

5.2 Building of the 3D virtual test bench in QuickSim

Concerning figure 5.1, the whole test bench was reproduced in the simulation environment of QuickSim.

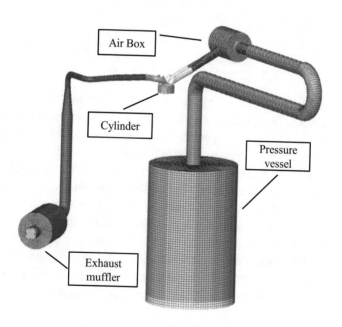

Figure 5.1: Reproduction of the test bench layout including pressure vessel and exhaust system [21].

The setting of simulation boundaries corresponding to environmental conditions allowed to minimize their influences on the resulting cylinder flow field. No manifold 1D-boundary-conditions were needed any more. Mainly, this allowed a detailed analysis of volumetric efficiency, residual gas concentration at intake valve closure (IVC), fuel mixture formation, turbulence profile and combustion progression [11]. Besides, the possibility to study the scavenging process, together with the combustion and the injection event was essential to reproduce the behaviour of the external water injection, particularly for IWI strategies. The investigation of the whole engine system though 3D-CFD simulations provides the possibility for the virtual engine design.

More specifically, as reported in [11], the models adopted for the calculation of the flow field in a full engine simulation are a combination of different approaches:

- traditional local 3D-CFD-models
- engine-specific phenomenological relationships
- trained neural networks
- databases
- empirical relationships

When considering the cell dimensions and structures, the models were specifically analysed to produce reliable results of engine thermodynamic process. Therefore an accurate prediction of the fuel heat-release and a realistic flame propagation was retained the main task of the combustion modelling, rather than detailed analysis of the chemistry involved in the process as described in detail in chapter 3.6. In addition, by linking in the software CFD-calculations and the *Real working-process analysis* (the same thermodynamic analysis adopted in indicating systems) it is possible to exclude any implausible differences when comparing heat release, wall heat transfer, internal energy variations, at each time-step [11].

Assumed so, a reliable mesh was built from the engine CAD including piston, cylinder head, combustion chamber, intake and exhaust manifold. Separately, the auxiliary systems of the test bench (pressure vessel, air-box and exhaust system) were measured and reproduced in the simulation environment, as shown in figure 5.1, being aware of the exact sensor positions, for a reliable comparison with measurements.

Nevertheless, the focus of the analysis is the investigation of the engine flow field with water injection and particularly its influence on the combustion process. For this purpose, the cylinder head was built in order to reproduce the water injector position at the test bench. The left side of figure 5.2 shows the injectors housed in the single-cylinder engine head, especially the manifold water injector (a) and the lateral direct water injector (b). Again these devices correspond to *Injector 1* and to *Injector 2* listed in table 4.1. The right side of figure 5.2 pictures the corresponding engine mesh built in QuickSim. Beyond position (a) and (b), it was possible to investigate virtually two additional injector positions, namely (c) corresponding to port water injection (IWI) and (d) representing a central direct water injector. These additional configurations were only investigated through the virtual test bench created in QuickSim, after calibration of the base water injection cases through the measurements.

Single-cylinder engine   QuickSim virtual test bench V10

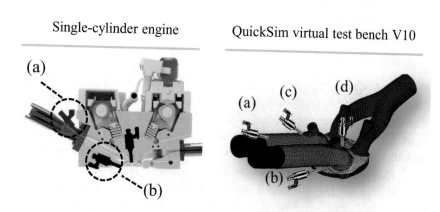

Figure 5.2: Injector layouts for water injection in the real engine (left) and in the 3D-CFD virtual test bench built in QuickSim (right) [48].

5.3 Model validation and calibration

In order to analyse the effects of water on combustion and engine parameters separately, tests with no water injection were compared to water injection cases, without changing the spark advance, meaning that the Knock sensor was not correcting the combustion phasing. Then, by injecting water, a new spark advance was found reaching again the Knock limit of the specific load point. Figure 5.3 presents this procedure, where a thermodynamic investigation was carried out primarily under Knock-limited calibration with stoichiometric mixture.

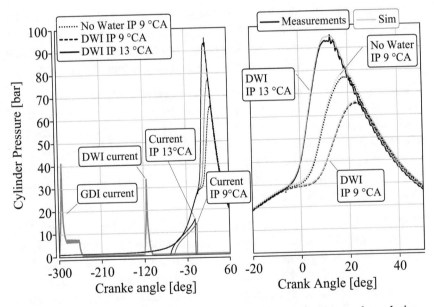

Figure 5.3: Effect of DWI on combustion at constant or advanced spark timing up to Knock limitation, at OP2 with 20 % W/F ratio and 100 bar injection pressure [48].

The selected operating point is OP2. Boost pressure was provided externally, and it was regulated to reach 1.3 bar absolute pressure, with wide opened throttle valve (WOT). A moderate knocking combustion is visible in the right side of figure 5.3, which represents a zoom of the corresponding left diagram, with a spark advance of 9 °CA b.FTDC. The injector current profiles, for both gasoline and the water, are represented in the left side of figure 5.3. In this case, water was injected directly in the cylinder at 100 bar and 120 °CA b.FTDC. By injecting water, combustion was initially slowed down, its centre moved towards late phasing, but the Knock was completely disappeared (dashed lines in figure 5.3). Furthermore, the consequence of adding water was a reduction of peak pressure of almost 15 bar, that could be recovered and even increased, by advancing the spark timing to 13 °CA b.FTDC and hitting again the Knock limit conditions (continuous lines in figure 5.3).

The experimental cylinder pressure traces of figure 5.3 represent one cycle over 500 cycles, where knocking combustion appears, this means that no pressure traces averaged was considered. Grey lines show the corresponding cylinder pressure traces from 3D-CFD simulations on the right side of figure 5.3.

The impact on engine parameters can be seen in figure 5.4, where DWI at 100 bar is reported for different W/F ratio, considering the procedure mentioned above. Following the Roman numerals in figure 5.4, I represents the base case without water injection, then II considers DWI at 20% W/F ratio and same spark advance of I, while III reports the results for an advanced spark time. Also, the W/F ratios was continuously increased at the test bench trying to improve the spark advance further.

At 20 % W/F ratio, injecting water, generated a reduction of T_3 by 10 °C, without spark advance modification. Water evaporates and influences combustion velocity, leading to an increase in combustion duration up to 6%. Then still considering 20 % W/F ratio, the combustion could be shifted to early phasing, by advancing the spark timing. The centre of combustion improved up to 8°CA a.FTDC, lowering as well the combustion duration with respect to the case without water injection and T_3 was reduced by 25 °C. The growth of water mass above 30 % W/F did not bring an enhancement of the Knock resistance, preventing any additional early shifting of spark advance. It was just experienced a reduction of T_3. This outcome was due to the combination of mostly two effects. On one side, higher amounts of water injected during compression phase were not able to evaporate because of high HOV, inhibiting water evaporation itself as discussed in chapter 1.2.3 and the further water evaporation took place at the end of combustion, with the only effect of reducing T3 (chapter 8 discusses this phenomenon). On the other side, higher water amount increased combustion duration and led to instability.

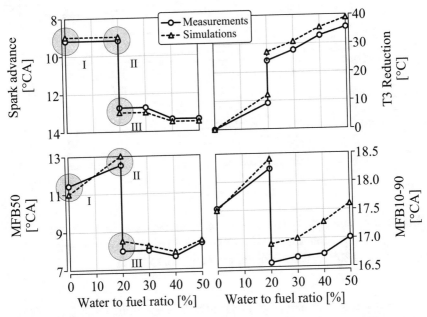

Figure 5.4: Combustion optimization procedure, at OP2 with DWI at 100 bar water injection pressure, in relation to different water to fuel ratio [48].

Regarding figure 5.5, a comparison between experiments with IWI (5 bar injection pressure) and DWI (100 bar injection pressure) is presented, for different W/F ratio. With the standard water injector engine layout, IWI showed higher potential for T_3 reduction, especially considering that DWI allowed earlier combustion phasing by higher Knock resistance with respect to IWI, which in turns led reducing as well T_3. The effect on T_3 for different water injection strategies, keeping constant the spark timing, will be investigated through simulations in chapter 8.

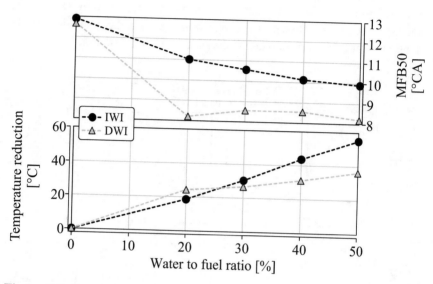

Figure 5.5: Comparison IWI at 5 bar injection pressure and 360°CA b.FTDC injection timing, with DWI at 100 bar injection pressure and 120°CA b.TDC injection timing, for OP2 [48].

An explanation of the more significant attitude in controlling Knock for DWI to IWI could be found with the support of 3D-CFD simulations, considering the temperature of unburned gas and the flame speed for the two cases as reported in figure 5.6. For this engine configuration at OP2, DWI reduces more temperature of unburned gas zones compared to IWI. Moreover, the flame speed proceeds much more rapidly during combustion for DWI, and the End-gas (final unburned gas during combustion propagation) is crossed by the flame front earlier than for IWI. In figure 5.6 the sudden temperature spikes show that the flame front reached the unburned gas and they turn into burned gas). The lateral DWI injector realizes a less homogeneous water vapour distribution in the combustion chamber to IWI, with lower influence on flame speed slow down. On the other side the better water homogenization produced by IWI, endorses a more gradual time and spatial evaporation, carrying the engine towards lower T_3 with respect to DWI despite the unfavourable combustion phasing.

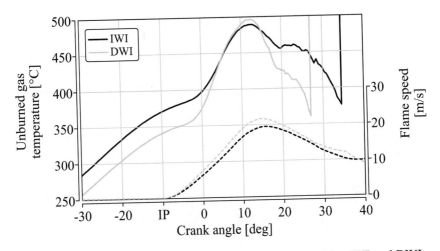

Figure 5.6: Temperature of unburned gas and flame speed for IWI and DWI case at OP2, 30% W/F ratio [48].

Since the benefit of water injection seemed to become weaker for W/F ratio higher than 30 %, and in order to propose a feasible approach also for water consumption, the further simulation analyses were carried out considering 30 % W/F ratio.

The decoupling of the effects of water on engine parameters allows calibrating the engine model in the 3D-CFD-Tool QuickSim, considering the data provided by the test bench for IWI and DWI. Then, the *Virtual test bench* studying the whole engine system (including air-box and exhaust system) was used to investigate new injection configurations and load points that would have been demanding to research with the current test bench configuration because of costs, time or hardware limits.

6 Water Injection for Enhancing Knock Resistance

As mentioned in chapter 5, OP1 (2000 rpm, 20 bar IMEP) presented a marked tendency to knocking combustion. Therefore the spark advance for OP1 was set to 1 °CA b.FTDC in the base engine calibration. Considering OP1, in the following analysis different parameters were varied, and their combination tested, to find the most promising strategies for the rise of the engine indicated efficiency while limiting at the same time water consumption. The investigated parameters were the start of water injection (SOI_W), the injection pressure, the injector position and the water to fuel ratio for indirect and direct water injection. Nevertheless in the next section, a selection of the most effective parameter variations is considered as resumed in table 6.1.

Table 6.1: Selection of the parameter variation for the water injection strategy adopted for experiments and simulations [49].

2000 rpm - 20 bar	IWI	DWI
Injection pressure [bar]	7/10/15	50/100/200
Start of water injection [°CA b.TDC]	360	120
Water injector position	Intake manifold (80 mm from cylinder head)	Lateral below exhaust valves
Injector holes	8	5
Injector targeting	Standard (water)	Standard (gasoline) / IVK

A range from 7 to 15 bar injection pressure was explored for IWI, while 50, 100 and 200 bar were tested for DWI. A fixed injection timing, for both IWI and DWI, is considered for this analysis. Also, just the injector positions according to the real test bench configuration are here presented for the Knock investigation (in chapter 8 the analysis will also focus on some different positions of the water injectors for rated power operations [48]). As a result, one single injector in the intake manifold was considered for IWI concepts (*Injector 1* reported in table 4.1). In contrast, a lateral direct injector, below the intake valves was investigated for DWI strategies (*Injector 2* reported in table 4.1). The intake manifold injector was a device optimized for water, featuring 8 holes and with jet angles directed 20 ° to the injector axis, towards the cylinder head. Specifically for DWI, the standard injector targeting from the gasoline device was tested with water ("Std."spray targeting), as well as with an injection pattern called "IVK". The last is a new water-developed spray targeting generated to improve water vaporization and its effect on the combustion process. The current analysis focuses on tests at 30 % water to fuel ratio. The calibration of the amount of injected water has to fulfil the optimal compromise between water consumption and exhaust gas temperature control (T_3) since the mass of injected water itself plays a fundamental role to reduce T_3. The rated power point of the engine at the test bench is investigated in chapter 8, with different water injection strategies in order to control T_3 with stoichiometric combustion [48]. Specifically in chapter 8, it is described how for rated power operation (4500 rpm, 20 bar IMEP), an optimal water injection strategy was already achieved by 30 % water to fuel ratio. This amount of water was found as the water needed to keep T_3 below the limit, having as reference the fuel enrichment case ($\lambda \simeq 0.85$).

Since it was also a good compromise in terms of water consumption, 30 % water to fuel ratio was set as maximum water injected quantity also for the operating point 2000 rpm 20 bar IMEP (OP1), investigated in the current chapter for knock resistance enhancement (the engine does not need fuel enrichment for component protection at this load point, see table 5.1).

The parameter variation was analysed comparing measurements and CFD simulations except for DWI case with IVK targeting that was just virtually investigated. The simulations were calibrated considering as acceptable 2 % maximum error between the following parameters:

- 50% Burned mass fraction (MFB50)

- Combustion duration corresponding to 10-90% burned mass fraction (MFB10-90)
- Indicated pressure (from low/high pressure signals)
- Air mass flow
- Fuel mass flow (and thus global λ)

With the aim to study the different impacts of water on the combustion, the procedure to couple measurements and simulations, presented in chapter 5.3 [48] was adopted as well for OP1. The operating point was run without water injection, and then water was introduced, but without modifying the spark advance, meaning that the Knock sensor was not correcting the combustion phasing. With the injection of water, the centre of combustion moves towards late phasing, and the Knock disappeared. This procedure was realized for IWI at 7 and 10 bar injection pressure, and for 50, 100 and 200 bar injection pressure for DWI cases, with standard injection targeting. By doing so, the load point was not reached anymore due to the water slowing down the combustion process (except when specific DWI strategies could increase turbulence and mixture homogenization which led improving the combustion phasing, before advancing the spark timing, as presented in chapter 6.2). In order to recover the load point, then a new spark advance was found by reaching the Knock limit again. By moving earlier the spark timing and checking the onset of Knock, in some cases it was possible to enhance the indicated efficiency of the engine, meaning that, by keeping the same injected fuel the torque was increased.

Such a methodology allowed to model the direct influence of water on the laminar flame speed, on the ignition delay and thus on the Knock occurrences (see again chapter 5.3).

6.1 Indirect water injection strategies

6.1.1 Influence of water injection pressure

Considering IWI for OP1, a single low-pressure injector was mounted in one of the two intake runners in the cylinder head, as mentioned before. This solution was found as the best compromise between costs and timing for the manufacture of the cylinder head. The range of injection pressure during the experiments varied between 7 and 10 bar, while the start of water injection was kept constant at 360 °CA b.FTDC, with reference back to table 6.1. Therefore the end of injection was changing accordingly to the respective injection pressure, but controlling as well the end of injection (EOI) terminated before IVC.

Several works such as [39] [48] highlighted the poor evaporation tendency of water during engine operations, especially for IWI strategies. They also described 3D-CFD simulations as essential means for studying the interaction of water with the engine flow field in order to control its evaporation. Bearing in mind the results in terms of reduced water evaporation obtained in [48], it was also virtually investigated a higher injection pressure of 15 bar to increase the water evaporation rate. Concerning figure 6.1, the analysis focuses on the water mass balance in the cylinder, during the compression stroke.

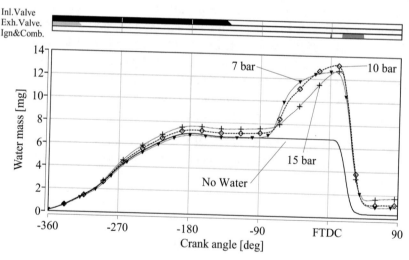

Figure 6.1: Crank angle-resolved evaporated water mass, for different injection pressure and intake manifold water injection strategies.

The reference test without water injection (continuous black line) is compared with three cases with water injection and different injection pressure (7/10/15 bar).

All CFD simulations were carried out by considering 70 % initial relative humidity of air (RH), meaning that moist air is used instead of dry air in the calculations. In order to understand the weight of considering relative humidity of the air in the simulations, it is useful to calculate the water vapour already present in the dry air mass at ignition point, if the sucked air is at 70 % RH. Injecting 30 % water to fuel ratio corresponds to almost 16 mg injected water mass at this load point (53 mg injected fuel), while 7 mg water is the contribution in terms of water vapour already present in the air at ignition point, due to consider moist air at 70 % RH instead of dry air. Despite that, water evaporation resulting from external water injection did not overcome 40 % of the injected water mass. The low evaporation tendency of water arises from its high surface tension, but mainly from its extremely high heat of vaporization. As a consequence, during the droplet's evaporation, a local fall of the mixture temperature prevents further water evaporation, by decreasing the saturation pressure, as already explained in chapter 1.2.3. Figure 6.1 also shows a comparable overall evaporated mass in the cylinder, at FTDC by changing the injection pressure, but with discrepancies in the evaporation process. Particularly, by 15 bar injection pressure regulation, the water travelled towards the cylinder faster, realizing an earlier and more progressively evaporation process, while for 7 and 10 bar injection pressure the rise of the evaporated mass occurred later and suddenly.

These differences in the evaporation process reflect directly on the cylinder temperature trend, as can be seen in figure 6.2. Here the temperature reduction up to FTDC is reported again for the case without water injection for different water injection pressures. Before ignition, injecting water at 7 bar led to a temperature reduction of 10 °C, while with 10 and 15 bar injection pressure the decrease in temperature was respectively 22 °C and 28 °C.

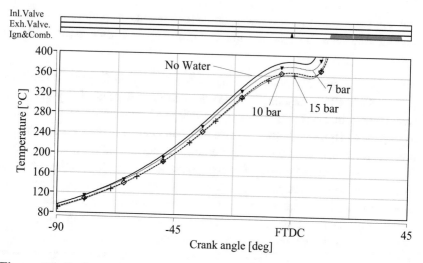

Figure 6.2: Reduction of the in-cylinder temperature at ignition point for different injection pressures IWI strategies.

Since at first the combustion parameters such as ignition point were kept constant, by injecting water a reduction of maximum in-cylinder pressure and pressure gradient occurred, as presented in figure 6.3, for different injection pressures. This effect arose from the reduction of in-cylinder temperature due to water evaporation, but also because the addition of water acted as inert mass and slowed down the laminar flame speed as demonstrated in chapter 3.5.

Nevertheless, considering one more time figure 6.3, the higher the water injection pressure, the lower the in-cylinder pressure reduction. These results seem to be in contrast with the reduction of in-cylinder temperature seen in figure 6.2, but as well with the reduction of the laminar flame speed, with respect to the increase of the water content shown in figure 3.3 in chapter 3.5. If the higher water injection pressure caused a reduction of in-cylinder temperature during the compression stroke, this should have produced a reduction of the peak pressure as well (at constant ignition point and lambda). However the reduction of the in-cylinder pressure realised during the compression stroke, was somehow compensated, and finally, the case at 15 bar water injection pressure showed the highest in-cylinder pressure peak (all the investigated water injection configurations correspond to 30 % W/F). Besides, the injection of water at 15 bar was responsible for slightly higher water evaporation which in turn should have reduced the pressure in the cylinder more dramatically to 7 bar injection pressure, since the higher available water vapour content should slow down the laminar flame speed more drastically. By analysing the laminar flame speed produced by the case at 15 bar water injection pressure, a lower trend came effectively upon as can be seen from figure 6.4. Higher injection pressure brought to higher evaporation rate and greater water vapour content, mitigating more the laminar flame speed (comparable results for 10 and 15 bar injection pressure). However, the resulting turbulent flame speed was higher for higher water injection pressure.

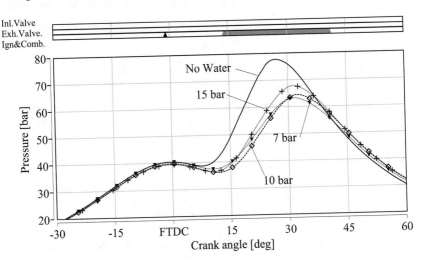

Figure 6.3: In-cylinder pressure variation for different IWI strategies and injection pressures, at constant ignition point.

The reasons for this opposite trend could be investigated through CFD simulations and particularly, by analysing the flow field induced by water injection. Specifically, different water injection pressures modified individually the turbulence and the mixture formation (and therefore fuel evaporation) within the cylinder. With reference to figure 6.5 and to figure 6.6, respectively the Tumble motion and the turbulent kinetics energy (TKE), for each water injection cases is reported. Particularly, figure 6.5 shows an enhanced Tumble by injecting water at 10 bar, and especially at 15 bar, with respect to the case at 7 bar. The water jets travelled the intake manifold faster and modified the flow field accordingly. Furthermore with reference to figure 6.6 the TKE in the cylinder is pictured (left side of figure 6.6), with respect to the one produced in a small volume surrounding the spark plug (right side of figure 6.6). At 10 and 15 bar injection pressure, the turbulence at the spark plug increases, just before the spark ignition and therefore, a faster combustion event was realised compared to 7 bar water injection pressure.

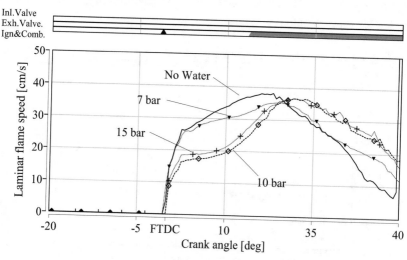

Figure 6.4: Laminar flame speed variation realized through IWI, at different injection pressures and without modifying the combustion parameters such as ignition point.

Moreover considering the mixture formation, the cases at 10 and 15 bar water injection pressure produced higher fuel evaporation, due to the slightly increased turbulence (up to 3 % more evaporated fuel at ignition point for the case at 15 bar injection pressure to the case at 7 bar). With reference to figure 6.7, almost no differences can be seen in the lambda distribution of the cylinder (left side of figure 6.7).

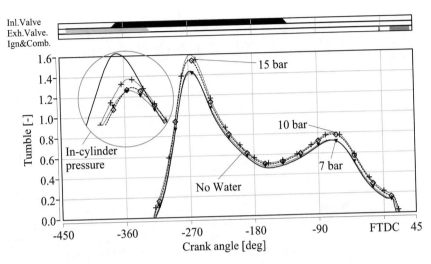

Figure 6.5: Modification of Tumble and of pressure peak (upper left), resulting from different IWI strategies.

Differently, in a volume surrounding the spark plug (right side of figure 6.7), the fuel evaporation for the case at 10 and 15 bar water injection pressure occurred earlier, and it realized a better mixture homogenization. At 10 and 15 bar water injection pressure the water reached the cylinder earlier compared to 7 bar injection pressure and the realized more gradual evaporation had a minor influence on fuel evaporation. The combination of these effects was responsible for the triggering of a faster combustion event for higher water injection pressure and therefore, for mitigating the effect of water in shifting the combustion towards late phasing.

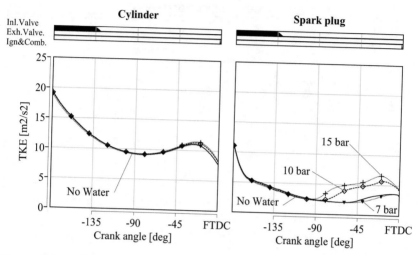

Figure 6.6: TKE resulting from different IWI strategies in the cylinder (left) and in a small volume surrounding the spark plug (right).

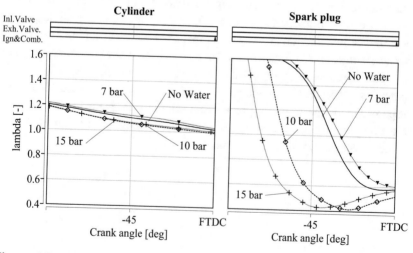

Figure 6.7: Air to fuel ratio from different IWI strategies in the cylinder (left) and in a small volume surrounding the spark plug (right).

6.1.2 Combustion and Knock analysis

The following investigation was realized with the support of the Knock model described in chapter 3.4. The model was calibrated based on the experimental data available from the single-cylinder engine test bench, and specifically for the current engine operating point. As mentioned before, the single-cylinder engine at the test bench was calibrated at the limit of knocking combustion for OP1, which corresponded to set the ignition point at 1 °CA b.FTDC. With this spark advance, through the comparison with experimental data and the simulated combustion parameter (in-cylinder pressure curve and burn rate), it was found that up to ~ 6.5 % of mass in self-ignition condition could be tolerated before severe Knock occurs at this load point. The results are shown in figure 6.8, where the top diagram represents the burned mass fraction, while the bottom one depicts the percentage of mass in self-ignition conditions, for each test. This last parameter serves as a Knock index. It represents an evaluation parameter coming from the simulation, which on one side is affected by some inaccuracies with respect for example to direct detection of Knock onset through a Knock sensor at the test bench, but on the other side it allows a quite effective relative comparison between the different water injection strategies in the simulations. It must be recognize that in the definition of the Knock index the weakest assumption done is to consider in the heat transfer model a constant value for the wall temperature. This assumption can generate inaccuracies in the absolute value of the mass in self-ignition condition, an error that is almost compensated when considering a relative comparison of similar water injection strategies.

More specifically, the limit before Knock occurrence at OP1, corresponds in the simulations to the maximum value assumed by the mass in a self-ignition condition of the case without water injection, which was calibrated at the test bench slightly before the onset of Knock. By adding water, the combustion process shifted towards late phasing and in-cylinder temperatures, and pressures decreased, therefore Knock tendency reduced, as showed by the lower mass in self-ignition conditions in figure 6.8 (bottom diagram). As a result of unchanged combustion parameter (constant spark advance), the centre of combustion freely moved from 21 to 27 °CA a.FTDC, when comparing the case without water, with the one at 7 bar water injection pressure. Since the slightly increased turbulence and improved mixture formation, the combustion process realized earlier for the case at 10 and 15 bar water injection pressurecompared to the case at 7 bar water injection pressure as explained in chapter 6.1.1.

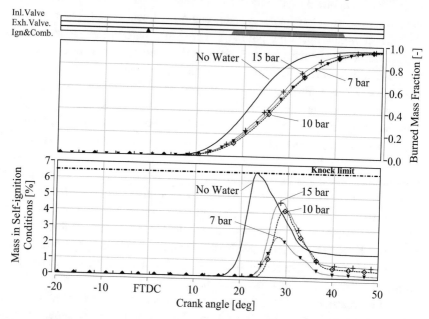

Figure 6.8: Burn mass fraction (top diagram) and mass in self-ingnition conditions (bottom diagram) for different water injection pressure
and constant ignition point (1°CA b.FTDC).

Certainly, if the centre of combustion is free to vary (without adjusting the ignition point the MFB50 is delayed by water injection), the load point, cannot be reached any more by injecting water. For example, with water injection at 7 bar, a reduction of 6 % of the torque was obtained. Considering figure 6.9, the ignition point was varied in order to achieve the target torque again, but respecting the Knock limit constraints (~ 6.5 % of the mass in self-ignition condition). For the different water injection pressures, it was found that 4 °CA b.FTDC was the ignition timing limit before knocking combustion, with minor differences between the water injection pressure cases. From figure 6.9, it is also evident that there were no margins for improving the combustion phasing

of the base calibration (without water injection), once considering 30 % water
to fuel ratio, at 360 °CA b.FTDC and with the injector in the intake manifold
position.

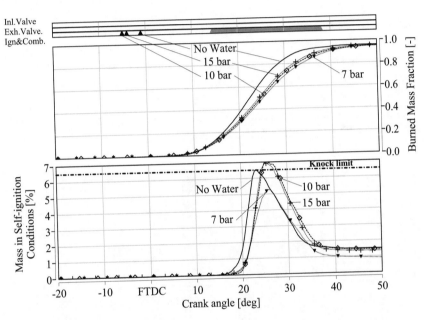

Figure 6.9: Burn mass fraction (top diagram) and mass in self-ingnition conditions (bottom diagram) for different water injection pressure
with respect to sweep of spark advance.

Figure 6.10 resumes conveniently the ignition point sweeps for each water injection pressure. The first data on the left of figure 6.10 represents the base calibration without water injection; moving to the right, the results for the different injection pressure are illustrated. With dashed lines and for each analysed
parameter on the y-axis (IMEP/MFB50/indicated efficiency), the reference values are extended to the right starting from the base calibration without water
injection. Minor differences in the setting of the ignition point were recorded
for different water injection pressures.

By injecting water at 7 bar, for example, MFB50 moved from 21 °CA after firing top dead center (a.FTDC) to almost 27 °CA a.FTDC, causing the indicated efficiency reducing from 30.5 % to 29 %. The only possibility to reach again the target torque and further increase the engine indicated efficiency to the reference case was to adjust the spark timing at 5 °CA b.FTDC as shown in figure 6.10 (7 bar injection pressure). Nevertheless, with such a spark advance, the Knock limit was overcome (11 % of the mass in self-ignition conditions) and thus the spark limit was consequently adjusted to 4 °CA b.FTDC. Concerning the case at 10 and 15 bar water injection pressure, the spark advance at the limit of knocking combustion was respectively 5 °CA b.FTDC and 4 °CA b.FTDC. Also, no margins were found for increasing the engine indicated efficiency with such water injection strategies.

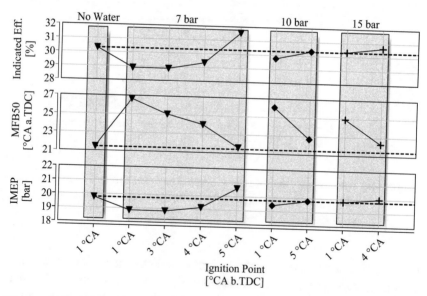

Figure 6.10: Comparison of indicated efficiency (top), MFB50 (middle) and IMEP (bottom) for different IWI strategies and injection pressures (grey frames), in relation to sweep of spark advance [49].

In figure 6.11 are reported in-cylinder temperature at ignition point, MFB10-90 and in-cylinder peak pressure for each water injection pressures, at different the spark advances. The injection of water increased combustion duration and for example, considering the 7 bar injection pressure case, it moved from 21 up to 30 °CA. The rise in the combustion duration for higher water injection pressure (10 and 15 bar) was less dramatically with respect to 7 bar injection pressure, since the support of a slightly enhanced turbulence, as seen in chapter 6.1.1.

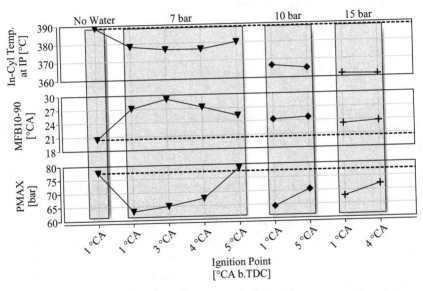

Figure 6.11: Comparison of in-cylinder temperature (top), MFB10-90 (middle) and PMAX (bottom) for IWI at different injection pressures (grey frames) in relation to sweep of spark advance [49].

New water injection configurations have to be considered to enhance the engine indicated efficiency of OP1. The possible further solutions could be to move the injector to the inlet port, for intensifying water evaporation, or raise the water to fuel ratio above 30% (changes in the injection timing had minor influences for Knock control purposes). These possibilities were not directly evaluated in the current work for OP1. However the modification of the indirect injector position is screened in chapter 8.1 at least for OP3 and T_3 control.

Since the objective was to compare IWI and DWI, by reducing at the same time the water consumption, just 30 % water to fuel ratio was investigated. Therefore, other strategies have to be evaluated to enhance the efficiency of OP1, with the current engine layout. As a result, DWI was considered keeping the same amount of injected water. As a matter of fact, DWI is expected to enhance the water evaporation rate and thus its cooling power to IWI [37], leading to higher Knock resistance.

6.2 Direct water injection strategies

This chapter provides an in-depth investigation of the possible DWI strategies and configurations. At first, the injector targeting of the DWI was virtually optimized to increase the air entrainment into the spray and to improve the water evaporation tendency. In addition, a new orientation of the direct water injector allowed injecting water in a direction coherent with the Tumble motion within the combustion chamber. The water injection in base configuration was directed towards the piston head, since the injector used was a standard gasoline device developed to improve the mixture formation during the entire intake and compression stroke. On the contrary, the water could be injected later and specifically at 120 °CA b.FTDC, thus quite delayed to fuel injection (380 °CA b.FTDC). With the injection timing mentioned above, the water injection could be used to support the Tumble and even increase the in-cylinder turbulence. It is then pointed out how the injection of water could be exploited directly to improve the combustion phasing, already before adjusting the ignition point.

6.2.1 Optimization of the water injector targeting

As discussed, the engine at the test bench was equipped with an additional water direct injector placed below the inlet valves, with a spray targeting developed for injecting gasoline in a different engine application (*Injector 2* with reference to table 4.1). The lateral injector housing was manufactured so that the injector axis was almost perpendicular to the spark plug. The injector targeting consists of a narrow jet with a characteristic angle directing the spray towards the piston head. The standard injector targeting corresponds to the grey spots in figure 6.12, where a perpendicular section to the injector axis, at 30 mm distance is considered. The targeting was measured through *Laser induced fluorescence (LIF)* in a constant volume chamber and resulted not symmetric. Since the low water evaporation was responsible for reducing the water cooling performance, a new injector targeting was developed to increase the air entrainment into the jets, trying to distribute the water enthalpy of vaporisation over a higher amount of mass. Figure 6.12 depicts the water-developed targeting with black spots, and in the next analysis, it is called IVK targeting (IVK) to recognise it from the Standard (Std.) one.

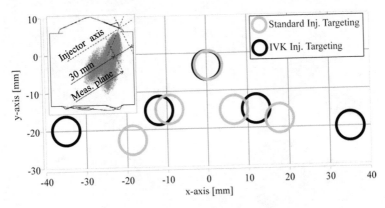

Figure 6.12: Standard gasoline injector targeting (Std.) versus water-developed injector targeting (IVK), at 30 mm section perpendicular to injector axis [49].

6.2.2 Turbulence enhancement

In addition to the distinctive spray targetings (IVK and Std.), two different angles for global spray pattern (with respect to injector axis) and two water injection pressures (100/200 bar) were investigated. The angle *Type 1* is the standard angle of the lateral gasoline device, injecting at 60 ° to the injector axis in the direction of the piston head. *Type 2* is the angle obtained by adjusting the spray towards the direction of Tumble motion, meaning 5 ° towards the piston head to the injector axis. The configuration with IVK injector targeting and angle *Type 2* is called IVK tumble targeting (IVK T). The directions of the spray are represented in figure 6.13.

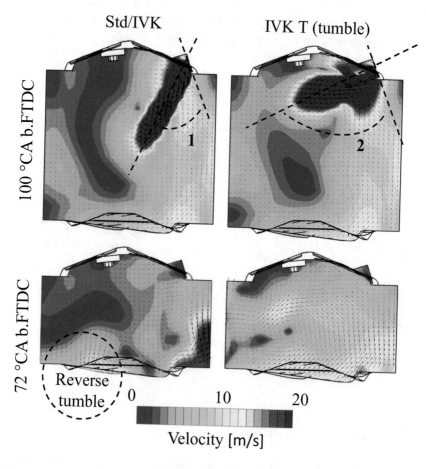

Figure 6.13: Comparison of Standard and IVK targeting with spray angle 1 (left), with respect to IVK targeting and spray angle 2 (right) at 100 °CA b.FTDC (top) and at 72°CA b.FTDC (bottom) [49].

On the left is displaced the Standard and IVK targeting with angle *Type 1*, while on the right side, IVK targeting is combined with angle *Type 2* to generate a turbulence-enhanced water spray targeting (IVK T).

Figure 6.13 shows the velocity flow field, through a section of the combustion chamber at 110 ° CA b.FTDC (on the upper part) and later at 72 °CA b.FTDC (on the lower part). Particularly, figure 6.13 presents a comparison between Std./IVK configuration (on the left side) with respect to IVK one (on the right side), considering 200 bar injection pressure. The angle *Type 2* supports the charge motion, while the angle *Type 1* produces a reverse Tumble which makes the forward Tumble weaker (see bottom left part of figure 6.13).

The matrix for the investigation of the spray patterns with DWI strategies is reported in table 6.2. For each test, it is specified the adopted injection pressure (100/200 bar), the angle type (*Type 1* for standard pattern spray angle and *Type 2* for turbulence-enhanced spray pattern) and a description of the targeting characteristics.

Table 6.2: Simulation matrix for the investigation of DWI targetings [49].

Test	Inj. Pressure [bar]	Angle type [-]	Targeting spec.
Standard (Std.)	200	1	standard targeting (narrow and direct towards piston centre)
Standard 100 (Std.100)	100	1	standard targeting (narrow and direct towards piston centre)
IVK targeting (IVK)	200	1	new targeting (wider and direct towards piston centre)
IVK Tumble targeting (IVK T)	100	2	new targeting (wider and direct along main Tumble)

The diagram in figure 6.14 shows the results of the tests with different DWI configurations, reported in table 6.2 and with respect to the base case without water injection (continuous black line), for OP1 (2000 rpm. 20 bar IMEP). Mainly, it presents the resulting in-cylinder pressures, without water injection, with DWI at 200 bar and Standard (Std.) targeting and the similar case at 100 bar injection pressure (Std.100), calibrated with test bench data. As said before, the investigation aimed at first to reproduce the water injection effect on the combustion process, before adjusting the combustion parameter such as ignition point. This means that with water injection the combustion process should shift at first towards late phasing (and it is almost always the case, except in some injection configuration as explained soon). Since the injected fuel quantity was kept constant, the target IMEP of the load point could not be reached any more with delayed combustion. This effect can be seen for the test at 200 bar and Standard (Std.) targeting, shown through dotted line and x-markers in figure 6.14.

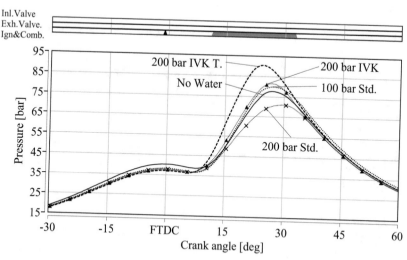

Figure 6.14: In-cylinder pressure variation at constant spark timing considering different DWI strategies and configurations, for OP1.

One interesting consequence of this approach consists in the possibility to analyse the direct influence of the water injection event on the combustion process.

Figure 6.15 shows the reason behind the water injection increasing in-cylinder pressure (before advancing the spark timing) presented in figure 6.14. Comparing 100 bar with 200 bar injection pressure (considering the same spray targeting (Std.) and 120°CA b.FTDC injection timing) it can be seen that the lower injection pressure caused a pressure rise instead of slowing the combustion event since the flow of the injected water at 100 bar injection pressure agrees with in-cylinder tumble (dotted line without markers). The resulting turbulence enhancement accelerates the flame speed propagation, and this effect overcomes as well the slowing down of combustion speed, due to water acting as diluting mass (see figure 3.3). The resulting in-cylinder pressure growths as a consequence of the turbulence rise, compensating as well the pressure reduction due to the water evaporating in the compression stroke (as it can be seen in figure 6.14 at FTDC).

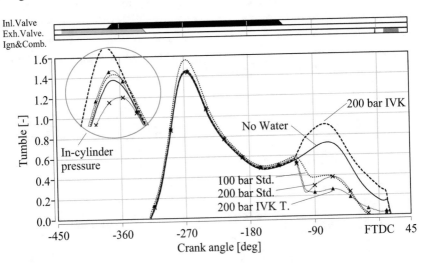

Figure 6.15: Modification of Tumble and of pressure peak (upper left), resulting from different DWI strategies, for OP1.

Since this effect was caused by the interaction between injected water and charge motion and depends on injection pressure, timing and targeting, it is clear that at first glance, it is not possible to assess whether a higher water injection pressure would achieve the best results. For this reason, the water injection strategy has to be investigated with respect to the load point and the injector layout, individually for each engine configurations. Again concerning figure 6.15, the dotted line with triangle-markers and the dashed line without markers respectively present the resulting in-cylinder pressure arising from IVK and IVK T water injection targeting. With a look to the upper left part of figure 6.15 (in-cylinder pressure peak), it is evident that the in-cylinder pressure enhancement of IVK T case was realised through higher turbulence and therefore this effect can be exploited to enhance the engine indicated efficiency long before advancing the spark timing.

Regarding figure 6.16, the left diagram presents the analysis of TKE in the cylinder for different DWI targetings, while the diagram on the right depicts TKE in a small volume surrounding the spark plug. The turbulence (TKE) comes from the dissipation of the Tumble when it is compressed, and therefore, the Tumble modification due to water injection reflected directly on the turbulence. Particularly all DWI strategies produced a rise of the turbulence in the cylinder (left side of figure 6.16), but just IVK T was able to increase it close to the spark plug (right diagram of figure 6.16).

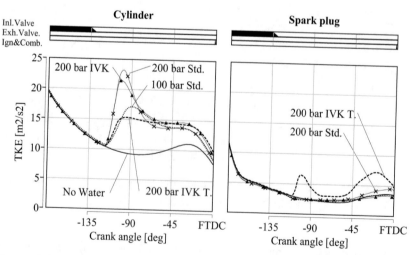

Figure 6.16: Modification of the TKE resulting from different DWI strategies in the cylinder (left) and in a small volume surrounding the spark plug (right), for OP1

For this reason, IVK T was responsible for a faster and shorter combustion event to reference case and the other DWI strategies. The overall evaluation of the different DWI targetings effectiveness needs to consider one more aspect. Particularly, the gradient of in-cylinder pressure rise is a function of turbulence and Tumble motion, but as well of mixture formation. As a consequence, the interaction between water and fuel during their evaporation have to be investigated.

6.2.3 Water evaporation and influence on mixture formation

Beyond the turbulence influence, is worth to evaluate the cooling power of the injected water, as it was done for IWI strategies (see figure 6.1). About figure 6.17, even though the overall injected water mass was the same as for IWI strategies, DWI allowed higher water evaporation rate at the ignition point (even with quite delayed injection). In this circumstance, the effect of higher injection pressure and the presence of a more favourable flow field for evaporation (cylinder pressure and temperature) helped the water passing into the vapour phase.

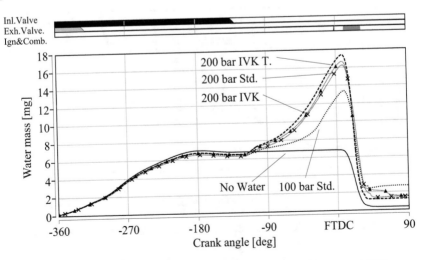

Figure 6.17: In-cylinder water mass balance considering different DWI strategies and configurations for OP1.

Almost 60 % of the injected water at the ignition point vaporised, with respect to 40 % evaporation rate of IWI strategy (15 bar injection pressure). However, a considerable amount of water vaporised during the combustion process or later, losing the cooling power potential of water for Knock resistance partially. The water evaporation could be improved thanks to valve timing strategies, for example, recirculating a higher amount of exhaust gas during the scavenging process, as mentioned later. Despite the limited evaporation, the water cooling power could be seen once the in-cylinder temperature at ignition point is analysed. Considering figure 6.18, the evolution of in-cylinder temperature is reported for the different injection strategies. Higher injection pressures brought lower in-cylinder temperatures, as expected, since the higher water evaporation. For example, by comparing the same injector targeting, 15 °C temperature reduction can be accomplished moving from 100 to 200 bar injection pressure, at ignition point. With regards to 200 bar injection pressure, IVK and IVK T targeting led to a further temperature reduction thanks to the broader spray, which was able to entrain a higher air mass between the jets.

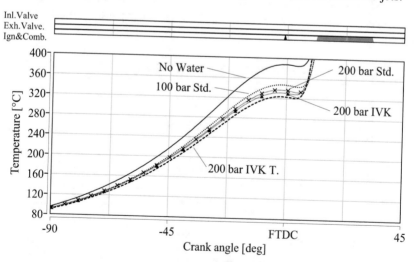

Figure 6.18: In-cylinder temperature variation considering different DWI strategies and configurations for OP1.

As a result, 20 °C temperature reduction was achieved moving from Std. targeting to IVK T. This step helps to spread the evaporation enthalpy of water on more mass and to avoid local temperature fall, which in turns would prevent further water droplets evaporation as presented in chapter 1.2.3. The IVK T configuration brought not only turbulence increase, but also higher water evaporation, as a result of both wider spray targeting and stronger turbulence. For this reason, IVK T targeting is also expected to be more effective in Knock mitigation.

By analogy with the analysis carried out for IWI in chapter 6.1.1, the study of the mixture formation was performed in relation to the different DWI configurations. The left diagram of figure 6.19 represents the spatial average of the lambda distribution in the cylinder, while the right diagram resumes the air to fuel ratio in a small volume surrounding the spark plug. Essentially, considering the whole cylinder volume (left diagram of figure 6.19), 100 bar water injection pressure caused higher fuel vaporization than 200 bar, generating a better mixture for Std. targeting. Concerning the right part of figure 6.19, IVK targeting presented delayed fuel evaporation, producing not homogeneous fuel mixture (dotted line with triangle-markers). At the same time, IVK T pattern generated anticipated fuel evaporation and a good mixture formation (dashed line without markers), beyond the turbulence increase.

These results could be explained by investigating the relation between fuel and water evaporation through CFD simulations. Fuel evaporation depends on charge motion but as well on the temperature field. Mainly, higher water evaporation decreases charge temperatures and thus prevents or slows down fuel evaporation. This effect can be highlighted by comparing the pattern Std. for both 100 and 200 bar injection pressure. The Std. targeting at 100 bar water injection pressure (Std.100) realized higher fuel evaporation with respect to 200 bar injection pressure since the reduced water evaporation at 100 bar injection pressure was cooling down the charge less remarkably than at 200 bar injection pressure (see again figure 6.18). As far as the quality of the mixture formation is concerned, it can be concluded that:

- Among the same spray pattern (Std. case) 100 bar water injection pressure resulted in better mixture formation with respect to 200 bar water injection pressure.

- On one hand IVK target enhanced water evaporation, but it caused also the worst fuel vaporization, which limited this pattern being effective for enhancing the engine indicated efficiency.
- IVK T pattern was responsible for the overall best mixture conditions, which brought to the trigger of a faster combustion, being the mixture conditions close to the spark plug more favourable (see right side of figure 6.19).

The effect of the water injection pressures for DWI configuration and the resulting fuel evaporation is further analysed for the spark advance sweep in chapter 6.2.7.

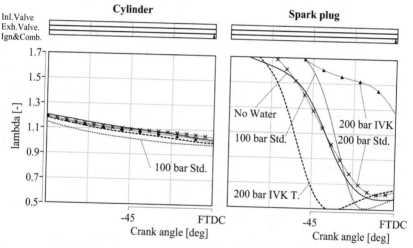

Figure 6.19: Mixture formation and resulting air to fuel ratio from different DWI strategies in the cylinder (left) and in a small volume surrounding the spark plug (right), for OP1

6.2.4 Combustion and Knock analysis

As shown in chapter 6.1.2 for IWI strategies, through experimental data it was found that up to ~ 6.5 % of the mass in self-ignition condition could be tolerated before knocking combustion occurs at OP1, by considering the numerical

Knock model. IWI shifted combustion towards late phasing, pressure and temperature reduced, thus Knock tendency dropped. Nevertheless, no margins appeared to boost combustion efficiency through advanced spark timings.

DWI instead, with optimized targeting was already able to improve the combustion phasing long-before shifting the ignition point and for specific strategies, as demonstrated in 6.2.2. Figure 6.20 shows that all DWI strategies decreased the Knock tendency and precisely the percentage of the mass in self-ignition condition reduced (bottom diagram of figure 6.20), to the no water case (continuous line). Besides, almost all the DWI pattern could shift the burned mass fraction towards early phasing, with exception to the Std. pattern (see the upper diagram of figure 6.20). Regarding the dashed line without markers, the burn rate analysis displays faster and more efficient combustion, especially for IVK T targeting.

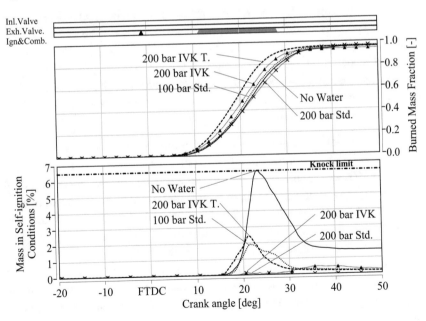

Figure 6.20: Burned mass fraction (top) and Knock evaluation (bottom), with different DWI strategies, at constant spark timing, for OP1.

For IVK injection configuration the centre of combustion improved but the combustion duration was more prolonged (see the dotted line with triangle-markers in the top diagram of figure 6.20), due to the less homogeneous mixture. The results showed above considered a constant spark advance for each water injection patterns, regardless of the different Knock sensitivity and thus with constant ignition point (1 °CA b.FTDC) and delayed centre of combustion (MFB50) when injecting water. For the different water injection patterns, the analysis focused on the corresponding effects on the engine flow field. In the next part, for each DWI strategy, the spark advance could be adjusted in order to reach again Knock limit conditions.

6.2.5 Results of the spark advance sweep

As presented in figure 6.21, each water injection pattern brought different spark advance and thus to different indicated efficiency gain. The different DWI patterns improved combustion phasing but differently and depending on the realized Knock resistance. With a sight to the top diagram of figure 6.21, the highest increase of the spark advance could be realized by Std. and IVK pattern at 200 bar injection pressure. With such a spark timing, the resulting MFB50 was close to ideal conditions (~ 8 °CA a.TDC).

A significant anticipation of the ignition timing was obtained as well with Std.100 and IVK T configurations, but hitting the Knock limit well before Std. and IVK water spray patterns. In these last two cases the combined effects of higher turbulence and more homogeneous mixture, respectively seen in chapter 6.2.2 and in chapter 6.2.3, were preventing further spark advance. Nevertheless, these two patterns and particularly IVK T one realized the conditions for a faster and shorter combustion event. For this reason, even though with IVK T, the spark advance could not reached such early value compared to Std. and IVK pattern, the effect of faster and shorter combustion event is expected to compensate the more delayed spark advance on combustion efficiency.

The next analysis addresses additional details of the influence of DWI injection pattern on the combustion, with respect to the spark advance sweep. Figures 6.22, 6.23 and 6.24 report the results of the ignition point sweeps. The diagrams show the modification of the engine parameters for different spark timings, up to reach the Knock limit again, and for different water injection configuration (grey frames). The dashed lines reproduce the reference value for each engine parameter, corresponding to the no water injection case (first grey frame on the left of each figure). The diagrams only report the results at 200 bar water injection pressure, for Std. (second frame from the left side), IVK (third frame from the left side) and IVK T (fourth frame from the left side) targetings.

The first test for each water injection strategy corresponds to the same ignition point of the no water injection case (1 °CA b.FTDC). Starting from figure 6.22 and considering at first the standard spray targeting, up to 15 °CA shift of the spark advance applied before reaching the Knock limit condition newly.

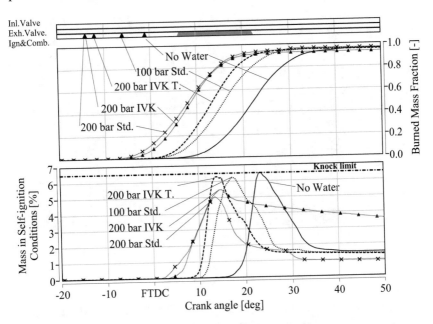

Figure 6.21: Burned mass fraction (top) and Knock evaluation (bottom), at different DWI strategies compared to the respective Knock-limit-spark advance, for OP1.

Therefore, the centre of combustion was accordingly advanced, but the maximum indicated efficiency improvement was accomplished at 13 °CA b.FTDC ignition point. At 13 °CA b.FTDC the optimal MFB50 was already reached and additional spark advance did not bring any further increase of indicated efficiency, due to the rise of heat transfer towards piston and cylinder wall. Despite the continuous maximum pressure increase (see figure 6.23), the work produced on the piston resulted lower, and thus no further improvement on indicated efficiency was achieved. Event tough OP1 was very sensitive to modifications of the spark advance, and since a considerable improvement of the combustion phasing was already realized at 8 °CA b.FTDC ignition point with DWI, the further advancement of the spark plug timing (from 8 to 16 °CA b.FTDC) produced a lower influence on the overall torque gain.

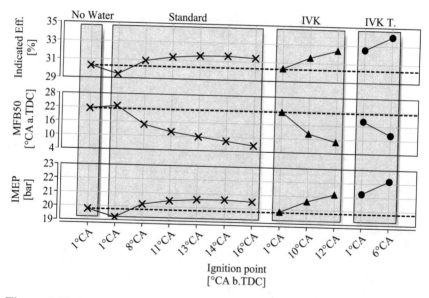

Figure 6.22: Indicated efficiency, MFB50 and IMEP for DWI with different injector targetings (grey frames), varying the spark advance, at OP1 [49].

Up to 12 °CA spark advance could be tolerated once using IVK targeting, but obtaining however higher indicated efficiency with respect to Standard targeting. For both Std. and IVK targeting, the MFB50 was improved up to 10 °CA a.FTDC enhancing the indicated efficiency to 31.7 % and 32.5 % respectively. Almost 34 % indicated efficiency was achieved with IVK T strategy, already with 6 °CA b.FTDC spark advance, which led to a centre of combustion of 13 °CA a.FTDC.

The highest efficiency rise obtained by IVK T configuration can be further analysed through figure 6.23. While the combustion duration (MFB10-90) was always between 19 °CA and 21°CA for the Std. water injection configuration, IVK and particularly IVK T targeting led to lowest combustion duration thanks to the higher turbulence. Besides, IVK T was also able to keep the maximum pressure below the engine limit (130 bar) with a wider safety margin in relation to other targetings.

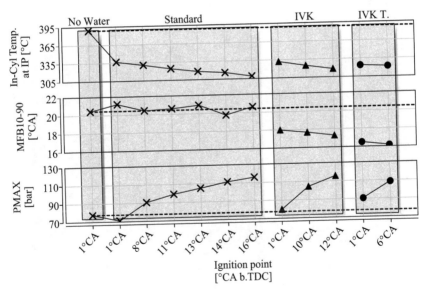

Figure 6.23: In-cylinder temperature, MFB10-90 and PMAX for DWI with different injector targetings (grey frames), varying the spark advance at OP1.

Figure 6.24 compares the vaporized fuel mass, the turbulent kinetic energy (TKE) and water evaporation rate for the different strategies, at the respective ignition point. IVK T was the only strategy producing sufficient overall fuel vaporization (see diagram on the top of figure 6.24) and consequently a highly homogenized mixture. In addition, the turbulence support through the optimized spray angle enhanced the water evaporation as well.

Finally, IVK T strategy raised the engine indicated efficiency up to 12 % to the no water injection case. The turbulence raise accomplished this outcome together with high water evaporation rate (and thus lower charge temperature) and homogeneous mixture. As a consequence, the centre of combustion shifted towards early phasing and the combustion duration was sensibly shortened. The final advanced ignition point was achieved thanks to the increased Knock resistance.

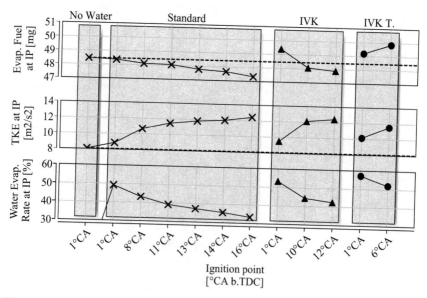

Figure 6.24: Evaporated fuel TKE and water evaporation rate, for DWI with different injector targetings (grey frames), varying the spark advance at OP1.

6.2.6 Mixture formation induced by water injection

The mixture formation and the homogenization can be deteriorated, improved and generally controlled depending on the direct water injection strategy used. In this paragraph the analysis done in chapter 6.2.3, for the mixture formation and the results illustrated in chapter 6.2.5, considering the advance of the spark timing are combined. Particularly, the IVK T targeting for DWI was developed with the purpose to support the fuel evaporation, providing the engine with a mixture that promotes the resistance to knock and triggers high combustion speed. Therefore, it is useful to have a look to the lambda map (see chapter 4.6) of the case without water injection (No Water) in comparison with the standard DWI (Std.) and with the IVK T strategy to examine the mixture formation at any crank angle before combustion.

With reference to figure 6.25 the Lambda map of the no-water case is sketched.

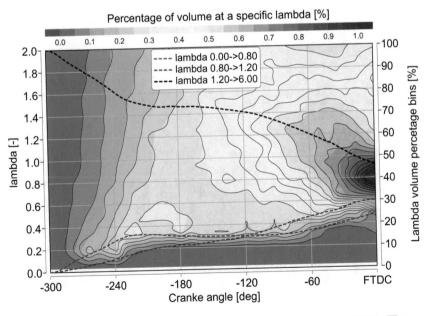

Figure 6.25: Lambda map for the case without water injection (OP1). The right y-axis shows the sum of the volume for each crank angle, grouped into different lambda ranges.

At each crank angle, the colour on the z-axis represents the percentage of the volume at a certain lambda value in relation to the total volume of the combustion chamber. The air-fuel ratio of each volume at each crank angle can be read on the left y-axis. The red zones highlight a larger percentage of the volume of a given lambda that corresponds to the left y-axis. Taking the right y-axis into account, the volumes are summed up to three different lambda intervals. The sum of rich volumes, where the lambda is between 0 and 0.8, is represented by the red dotted line; the sum of the volume in an optimized lambda interval is represented by the blue dotted line, while the black dotted line summarizes the larger volumes. Figure 6.25 shows that at this load point the engine has a non-optimized mixture with 30 % of the volume at FTDC with a lambda value close to 1 without water injection. Figure 6.26 shows the Lambda map of the case with non-optimized water injection (Std.) at 200 bar water injection pressure.

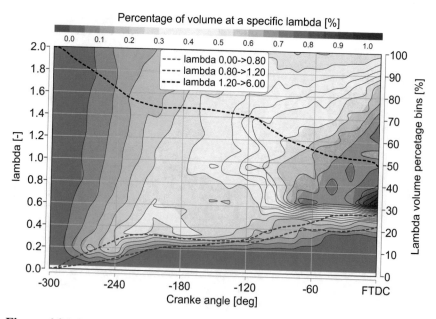

Figure 6.26: Lambda map for the case with Std. water injection (OP1). The right y-axis shows the sum of the volume for each crank angle, grouped into different lambda ranges.

Water evaporation reduces the quality of mixture formation by slowing down fuel evaporation. At FTDC, the volume composition is wider and more scattered in relation to the stoichiometric value. The lambda map of the IVK T. case is shown in figure 6.27 instead. The optimization of the water injection spray patter allows to increase the turbulence in the cylinder during the compression phase and in addition, the delayed water injection does not disturb the fuel evaporation by further lowering the temperature. As can be seen from the Lambda map of figure 6.27, the volume fraction at FTDC is close to the stoichiometric mixture value. The dashed blue line in figure 6.27 highlights that at ignition point the volume fraction in the ideal lambda range increases up to 40 %, thus improving the mixture homogenization and the knock resistance.

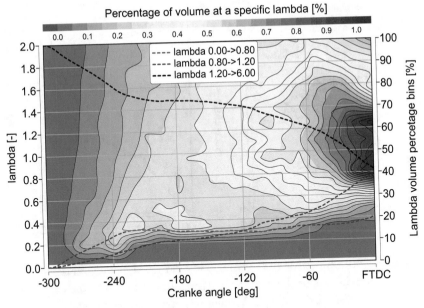

Figure 6.27: Lambda map for the case with IVK T water injection (OP1). The right y-axis shows the sum of the volume for each crank angle, grouped into different lambda ranges.

A direct comparison of the sum of the volume fractions in the rich and ideal lambda interval is also shown in figure 6.28 in the left and right diagrams. As can be seen from the curve in the left diagram, the Std. (DWI) case shows an increase in the rich volumes after water injection, which reduces the homogenization of the mixture. On the other hand, taking into account the right diagram of figure 6.28, the IVK T strategy produces an increasing gradient of the ideal lambda region exactly from the time of water injection. This analysis can also be used as a first evaluation of the emissions in a relative comparison with different water injection strategies. The IVK T strategy should have a lower soot production due to the higher degree of homogenization of the mixture. In a future work the soot production will be investigated by comparing the experiments with a new 3D soot model.

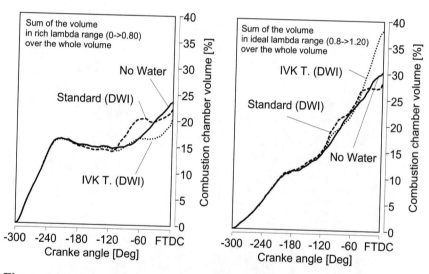

Figure 6.28: Comparison of the volume in a rich lambda interval (left) and in an ideal lambda range (right) for the case without water, with Std. and with IVK T water injection strategies (OP1).

The representation of the 3D flow field of the engine highlights the differences of the three analysed strategies considered in this paragraph, in relation to knocking behaviour. Figure 6.29 shows the percentage of unburned fuel mass (left side) in a section of the combustion chamber, compared to the local auto-ignition integral (right side and top view) considering the three investigated strategies. The comparison is performed taking into account the respective combustion centre (MFB50) for each engine configuration to compare the knock resistance at the same combustion time, even if the simulations have different ignition timing and combustion duration (MFB10-90).

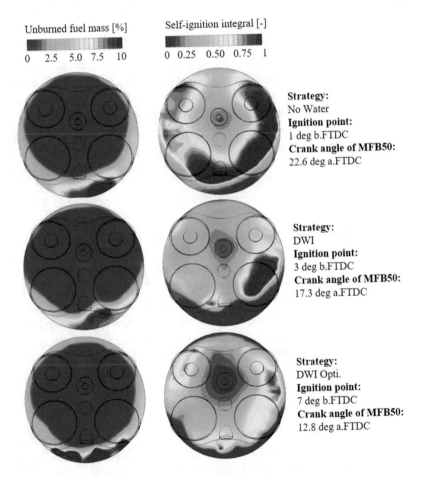

Figure 6.29: Comparison of the unburned gas mass and the auto-ignition integral (OP1). The knock event occurs when a certain amount of unburned gas mass reaches the state for self-ignition.

Again with reference to figure 6.29, if a certain mass of unburned gas is present and the auto-ignition integral reaches the value 1, the conditions for auto-ignition of the unburned gas are present. Nevertheless, a knocking event is only detected if a certain amount of unburned gas mass is present under these conditions, considering in this way the whole 3D flow field. Figure 6.29 still shows how the three operating points have a completely different flow field at centre of combustion. When viewed from above, looking at the left side of figure 6.29, the mixture formation near the wall seems to become more inhomogeneous, but this is mainly due to the fact that the considered MFB50 takes place earlier for IVK T case. Taking into account the respective MFB50, the IVK T strategy shows a less critical flow field for Knock onset, because the auto-ignition integral is smaller than 1 in the area where unburned gas are still present. The IVK T strategy results in high anti-knock performance, mainly due to the decrease in temperature of the unburned gas thanks to water evaporation.

6.2.7 Influence of injection pressure on spark advance

In chapter 6.2.2 the comparison between 100 bar and 200 bar water injection pressure (considering the same spray targeting (Std.)) demonstrated that at 120 °CA b.FTDC injection timing, the lower injection pressure was responsible for the in-cylinder pressure rise, instead of slowing down the combustion event. The reason was found in the in-cylinder turbulence agreement with the perturbation introduced by injecting water at 100 bar injection pressure. It was explained then that, at first glimpse, it would not be possible to assess whether a higher water injection pressure would lead to an increase of water injection performance.

This paragraph deals with the interaction between injected water and charge motion, once a sweep of spark timing is considered. Figure 6.30 depicts from top to bottom indicated efficiency, MFB50 and IMEP for three different DWI pattern. The first grey frame represents as always the reference case without water injection. The middle frame resumes the spark advance sweep done for Std. targeting at 200 bar shown also in figure 6.22, 6.23 and 6.24. The last frame on the right of figure 6.30 instead compare the spark advance sweep of the standard DWI patter but at 100 bar (Std.100). In agreement with the outcomes of chapter 6.2.4, lower injection pressure realized for Std. targeting higher indicated efficiency increase even though MFB50 could not be as much improved as for 200 bar water injection pressure case.

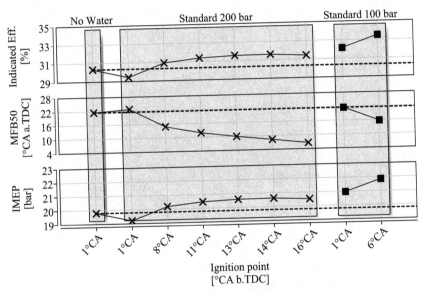

Figure 6.30: Indicated efficiency, MFB50 and IMEP at different injection pressures (grey frames) and same targeting (Std.), varying the spark advance, at OP1.

As mentioned in chapter 6.2.3, fuel evaporation and thus mixture homogenization have to be examined in combination with turbulence analysis in order to explain the resulting combustion process. This consideration can be confirmed once figure 6.31 is observed. Figure 6.31 indeed presents from top to bottom side, in-cylinder temperature, MFB10-90 and PMAX. Even though injecting water at 200 bar could lead to an early centre of combustion (see MFB50 in figure 6.30), the combustion duration (MFB10-90), when injecting water at 100 bar was shorter. Therefore, the combustion process was expected to be more efficient and less sensible to Knock occurrences, by compensating the effect of higher in-cylinder temperature (see again figure 6.31) to the case at 200 bar water injection, where the enhanced water evaporation brought stronger cooling of the mixture.

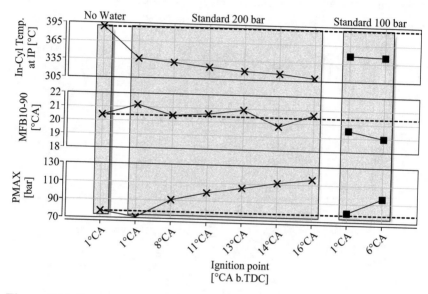

Figure 6.31: In-cylinder temperature, MFB10-90 and PMAX at different injection pressures (grey frames) and same targeting (Std.), varying the spark advance, for OP1.

Finally, figure 6.32 presents from top to bottom side, evaporated fuel, TKE and water evaporation rate at ignition point. On one side, lower water injection pressure decreased the evaporation rate of water and the charge cooling effect, but the accomplishment of fuel evaporation was slower and prevented by water injection. Therefore, higher fuel evaporation brought more efficient combustion.

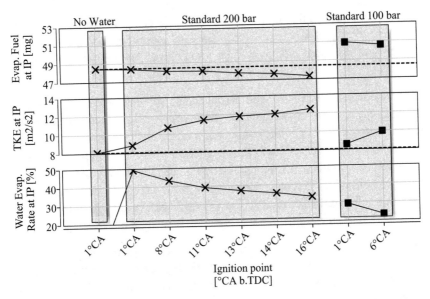

Figure 6.32: Evaporated fuel, TKE and water evaporation rate, at different injection pressures (grey frames), varying the spark advance, at OP1.

6.3 Water injection in combination with Miller cycle

This chapter investigates the possibility to enhance the engine compression ratio for OP1, which was already highly Knock limited and thus calibrated with delayed spark advance (1 °CA b.FTDC). The higher compression ratio demanded extraordinary Knock resistance, and thus new valve profiles were tested with Miller strategy. Since OP1 was the most critical in the engine map for Knock onset, it was worth to investigate the possibility of increasing the compression ratio on that point (worse case).

The challenge was to find a Miller profile that could provide a reduction of cylinder gas temperature, for such a low engine speed, keeping as well a proper mixture formation and homogenization. Therefore, two different valve strategies were selected and here analysed. The final assessment was the addition of direct water injection to the new valve strategies to increase engine indicated efficiency further.

Miller cycle can be realised with earlier Intake valve closure (IVC). As a direct consequence, the thermodynamic compression ratio reduces, and therefore temperatures at the end of the compression phase are lower, enhancing Knock resistance. Since the reduced Knock sensitivity, the possibility of a higher geometrical compression ratio can be exploited. The higher the geometrical compression ratio, the greater the effective work during the expansion stroke. Therefore, the combination of the Miller cycle and water injection can additionally lead to an increment in the engine indicated efficiency.

However, the advanced IVC brings a collapse of charge motion into the cylinder, reducing combustion stability, especially at high load. It is then somehow mandatory to adopt solutions to increase turbulence within engines if a Miller cycle is used. For example, a Tumble flap can be introduced, or new runners and piston geometry could be as well designed, for this purposes. A second possibility is to exploit direct water injection for increasing turbulence within the combustion chamber.

6.3.1 Increase of back-pressure

The measurements carried on the single-cylinder engine considered one bar back-pressure due to system constraints. The consequent simulation calibration was run with the same back-pressure because this limitation did not influence the relative comparison between different water injection strategies.

With the purpose to test the effect of different valve profiles and timings, it was worth to consider the real engine back-pressure realized at OP1. A value of 1.6 bar absolute back-pressure occurs in the corresponding multi-cylinder application, including after-treatment. The rise of the back-pressure to such a value reduces the effectiveness of the scavenging process, recirculating into the combustion chamber higher amount of exhaust gas. At the ignition point, the percentage of the exhaust mass trapped grew from 0.71 % to 1.43 % (with respect to fresh air) once the back-pressure increments. Nevertheless, the equivalent multi-cylinder engine tolerates this amount of exhaust gas, and it does not show deterioration on the combustion stability and any increment of the cycle-to-cycle variations.

Since it is intended to run the engine at $\lambda = 1$, it was necessary to increment the boost pressure and furthermore to adjust the ignition point for the measurements at lower back-pressure. As a matter of fact, the higher exhaust gas trapped into the cylinder and the increased boost pressure grew the temperature of the mixture and therefore the sensitivity to Knock became higher. The assessment on the ignition point adjustment was accomplished using the baseline data coming from the test bench and transferring the results for the new flow field condition using the virtual test bench. Although the delayed spark timing in case of the high back-pressure, the engine indicated efficiency improved because the temperature growth led to better fuel evaporation and homogenization at ignition point. Besides, lower combustion duration (MFB10-90) also confirmed this trend.

A further advantage to adopt higher back-pressure was found considering water injection. Regarding IVK T, at 200 bar injection pressure (DWI), the water tendency to evaporate before the ignition raised concerning the corresponding lower back-pressure case. This effect was due to the higher boost level and the more important exhaust gas trapped into the combustion chamber. The water evaporation rate grew from 55 % to 66 % at ignition point (considering the respective ignition point) by moving from 1 bar to 1.6 bar back-pressure. Thus the cooling power of water increased and compensated the higher charge temperature due to the rise of exhaust gas trapping.

The calibration of OP1 for each back-pressure cases, with and without water injection is reported in table 6.3. *Test 1* and *Test 2* correspond respectively to the " No water " case and to IVK T case investigated in chapter 6.2.4. In table 6.3 it is shown how the optimized water injection strategy improved the indicated efficiency up to 12 % (*Test 2* compared to *Test 1* at one bar back-pressure) with respect to reference case. This result was achieved by exploiting the turbulence increase generated by injecting water with optimized targeting directly into the cylinder, enhancing water evaporation and thus its cooling power (as seen in chapter 6.2.4).

A comparable enhancement of indicated efficiency was as well obtained for the higher back-pressure, between the case without water injection (*Test 3*) and IVK T one (*Test 4*). Table 6.3 also reports in-cylinder temperatures at 40 °CA b.FTDC. The temperature reduction was related to the different water evaporation rate, which influenced the Knock resistance as well. Almost 30 °C temperature reduction was possible with water injection, to the base case, for both back-pressure configurations. The derived turbulent flame speed included the effect of decreased laminar flame speed, due to water injection and the resulting flow field, but as well the turbulence increase caused by the direct water injection.

Table 6.3: Optimization of the ignition point with respect to the rise of p_3 and DWI.

Test	p_3 [bar]	W/F ratio [%]	Ign. Pnt. [°CA b.FTDC]	MFB50 [°CA b.FTDC]	Temp. at 40°CA b.FTDC [°C]	Indic. Eff. [%]
1	1.0	0	1	21.4	237	30.4
2	1.0	30	6	12.5	209	34.1
3	1.6	0	-3	24.2	240	30.9
4	1.6	30	6	11.6	215	34.5

6.3.2 Valve strategies and rise of the compression ratio

The influence of higher back-pressure was shown in the previous paragraph, compared to ambient back-pressure condition. The current chapter provides an insight on the new valve profiles selected, considering 1.6 bar back-pressure. Both the new valve profiles aimed to reduce the trapped air mass into the cylinder with early intake valve closure (EIVC). Especially the trapped charge undergoes an expansion before being recompressed during the compression stroke, decreasing charge temperature. EIVC is usually associated with the Miller cycle, and valve closure before the induction stroke is completed [58]. Also, the maximum valve lift is restricted by a shorter valve opening duration because of a limitation on maximum valve train acceleration. However, both a reduced valve lift and EIVC act to lower trapped charge [58].

Concerning these constraints, two valve profiles were generated, which differ just for the timing, as reported in table 6.4 and figure 6.33 (bottom diagram and right y-axis). M1 profile anticipates inlet valve closure by reducing valve opening window and maximum lift, but keeping the same overlap between intake and exhaust valves. M1 thus represents a classic *millerization* with quite reduced intake valve stroke (132 °CA opening width at 1 mm lift opening/closure stroke). On the other hand, M2 strategy adopts the same valve profile of M1, but with increased valve overlap, meaning with early intake valve opening (EIVO) with respect to M1. The target of further anticipating intake valve stroke was to exploit the first expansion phase to cool down the charge before compressing it [36]. In addition, the valve overlap extension allowed improving the scavenging process and reducing the trapped exhaust gas. This setting in turn results in the cylinder temperature reduction and therefore, in lower Knock sensitivity. Once water injection is considered, lower exhaust gas recirculation and lower boost pressure also led to reduce the water evaporation rate, being temperature conditions less favourable to water evaporation. M2 profile was chosen as a good compromise between cylinder scavenging and water evaporation enhancement.

Table 6.4: Settings of the investigated valve profile with Miller cycle.

Valve strategy	Max lift [mm]	Valve overlap [°CA]	Int./Exh. valve event length at 1 mm lift [°CA]	IVO [°CA b.FTDC]	IVC [°CA b.FTDC]
Base	10.15	51	194	389	120
M1	6	51	132	389	174
M2	6	80	132	408	203

In the following analysis, different simulations were compared by separating the effects of valve strategies, the rise of compression ratio and water injection. Other work such as Hoppe et al. showed an engine efficiency improvement with water injection by increasing the compression ratio by 8% [26]. In this case the engine compression ratio raised from 10.75 : 1 to 12 : 1 (corresponding to 12% rise). The modification of ε was accomplished by modifying the position of top dead centre, meaning that the cylinder head and piston were kept unchanged. The tested configurations are described in table 6.5.

As a matter of fact, *millerization* decreased engine volumetric efficiency and thus higher boost pressures were required to preserve the load point. The three aforementioned valve strategies (Base, M1 and M2) were adopted by adapting the boost pressure in order to keep the engine running at $\lambda = 1$ since reduced intake valve opening and higher compression ratio. The ignition points for each configuration were adjusted in order to reach again Knock limit.

Table 6.5: Configurations tested for different valve strategies, compression ratio and DWI.

Test	Valve strategy	Boost pressure [bar]	Ign. Pnt. [°CA b.FTDC]	ε [-]	W/F ratio [%]
3	Base	1.90	-3	10.75	0
4	Base	1.90	6	10.75	30
5	M1	2.06	-3	10.75	0
6	M1	2.10	-5	12	0
7	M1	2.10	-1	12	30
8	M2	2.16	6	10.75	0
9	M2	2.25	3	12	0
10	M2	2.25	5	12	30

Considering tests 3, 5 and 8 concerning table 6.5, the valve strategy M2 demands for higher boost pressure to the Base and the M1 ones since the scavenging process pushed fresh air towards exhaust line and less air resulted as trapped in the cylinder at IVC, as it is shown in figure 6.33. Both M1 and M2 avoided any back-flow from cylinder to intake system as instead happened for Base strategy (visible in figure 6.33). The absence of any back-flow was considered as an essential condition for the choice of the valve strategy since it allowed sensibly to reduce engine raw emissions. On the other side, the scavenging process was checked with respect to the value measured by the lambda probe, since air to fuel ratio measurements could be influenced by the air directly pushed towards the exhaust system.

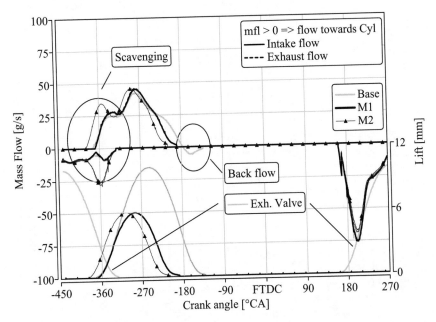

Figure 6.33: Flow realized between valves and cylinder for base, M1 and M2 valve strategies, for OP1 [49].

The main combustion parameters resulting from the different tests are summarized in figure 6.34. The diagrams show three blocks of tests from left to right. The first block on the left represents the results for Base valve profile, without and with water injection (respectively *Test 3* and *Test 4* considering x-axis of figure 6.34).The intermediate block presents the result for M1 valve strategies with $\varepsilon = 10.75$ and without water, then with increased compression ratio ($\varepsilon = 12$) and finally with increased compression ratio and water injection (respectively Test 5, 6 and 7 considering x-axis of figure 6.34).

The right block resumes the results for M2 profile with the same configuration order adopted for M1 analysis (*Test 8* $\rightarrow \varepsilon = 10.75$ - Test 9 $\rightarrow \varepsilon = 12$ - *Test 10* $\rightarrow \varepsilon = 12$ /DWI).

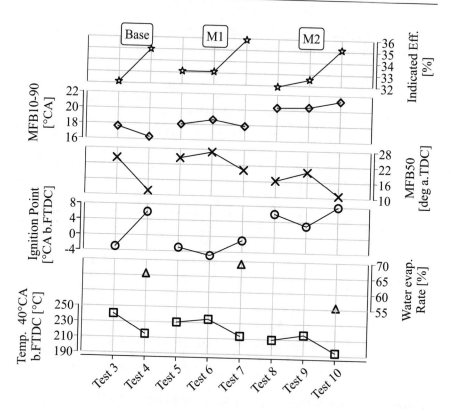

Figure 6.34: Results for Base, M1 and M2 valve strategies with higher compression ratio (Test 6-7-8-9), and water injection (Test 4-7-10), for OP1 [49].

Regarding figure 6.34, it is possible to compare again tests 3, 5 and 8, without water injection and with base compression ratio ($\varepsilon = 10.75$). Considering M2, the higher pressure difference between intake manifold and cylinder, once the intake valve was opening (lower trapped exhaust gas with respect to Base and M1), was responsible for expanding the charge and thus cooling it. In addition, in the M2 configuration, the intake valve closes when intake stroke is not yet finished (before compression), thus the charge was further cooled down. At 40 °CA b.FTDC the in-cylinder temperature achieved by adopting the Base profile (*Test 3*) was 240 °C, while for M1(*Test 5*) and M2 (*Test 8*) profile respectively

230 °C and 210 °C, without water injection (see first diagram from the bottom of figure 6.34). This outcome in turn means higher Knock resistance, thus the spark timing could be advanced up to 6 °CA b.FTDC for M2 case (see ignition point diagram of figure 6.34). No margins were highlighted instead with M1 configuration for advancing the spark timing since the higher trapped gas content to M1 let the engine being more sensitivity to Knock.

As consequence MFB50 of *Test 8* (M2) could be improved up to 16 °CA a.FTDC with respect to 24°CA a.FTDC of both *Test 3* (Base) and *Test 5* (M1), as it is presented in figure 6.34. Nevertheless, the extreme EIVC, resulting from M2 strategy caused a huge decay of in-cylinder turbulence and the combustion duration (MFB10-90) became longer (see again figure 6.34).

Besides, M2 also brought unfavourable fuel evaporation and thus homogenization as it can be seen in the 3D plot of figure 6.35. Here a comparison of mixture homogenization is realized at ignition point for Test 3, 5 and 8. *Test 3* had a sufficient final fuel stratification close to the spark plug, but the overall fuel distribution during the compression stroke was less uniform due to the back-flow mentioned above. Close to the spark plug, considering *Test 8*, the mixture was not any more well developed (even though the global in-cylinder mixture was always kept at $\lambda = 1$), where a too lean mixture at ignition point caused a slower combustion process. M1 valve profile was a good compromise in terms of in-cylinder temperature reduction, turbulence and mixture formation.

Besides the new valve strategies (M1 and M2), Test 6 and 9 had increased compression ratio ($\varepsilon = 12$), and the results are again shown in figure 6.34. The selected water injection pattern was the one with the rotated injector and broader spray targeting, in order to increase respectively turbulence and water evaporation (turbulence-induced water injection strategy called IVK T and investigated in chapter 6.2.1). The objective was the compensation for the turbulence loss in the cylinder due to the Miller strategy, with DWI support. The results are reported in figure 6.36. The diagram shows the Tumble motion of the Base valve profile with respect to M1 and M2, without water injection (continuous lines) and with direct water injection (dashed or dotted lines). As mentioned before M1 and M2 decreased the turbulence in the cylinder (M2 higher reduction to M1), but the injection of water tried to compensate the effect of EIVC by enhancing the turbulence before ignition point.

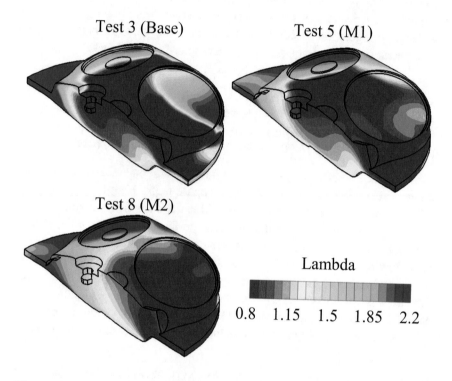

Figure 6.35: Fuel mass evaporated up to ignition point for the different valve strategies without water injection and with compression ratio 10:75, for OP1 [49].

With reference back to figure 6.34, Tests 4, 7 and 10 present the results with water injection respectively of Base, M1 and M2 valve strategies. Concerning figure 6.34, the water evaporation resulting from the M2 profile was lower because of intensified scavenging process, reducing the amount of trapped gas and therefore, the temperature of the mixture. This effect, in combination with the lower turbulence, prevented the water evaporation and reduced its cooling power.

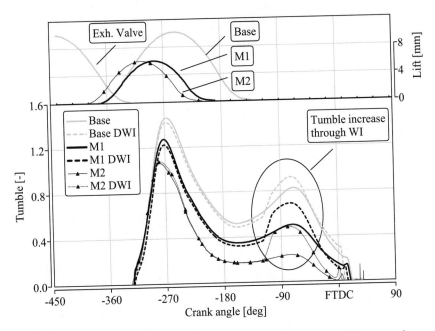

Figure 6.36: Turbulence increase through DWI ($\varepsilon = 12$) for different valve timings compared to the no water injection case ($\varepsilon = 10.75$), for OP1 [49].

Since the lower turbulence level and the addition of water slowed down the flame speed, M2 with direct water injection brought a further increase of combustion duration (21°CA as can be seen in figure 6.34).

On the other side, M1 with adequate homogenization and turbulence enhancement realised by DWI, in combination with high Knock resistance, due to raised water evaporation rate and lower in-cylinder temperature led to the most interesting results for the combustion process. When comparing *Test 3* (Base without water injection) with *Test 7* (M1 and DWI) the centre of combustion could be advanced of 4°CA, realising a comparable combustion duration, but adopting a higher compression ratio. This calibration helps the indicated efficiency rising from 30.9 % to 35.4 %, meaning 14 % consumption reduction (concerning the top diagram of figure 6.34).

7 Influence of Water Injection on Soot Formation

The current chapter reports an extract of the paper [21] and particularly the influence of water injection timing on the soot production, both considering experiments and 3D-CFD simulations. The emission investigations have been led at OP2, the only considered load point that did not present constraints neither in terms of Knock onset nor considering exhaust gas temperature before turbine. For such reasons the water injection potentials are investigate on OP2 not with the purpose of increasing the engine efficiency, but for evaluate the effects in terms of emission reduction.

7.1 Experimental results

Considering the single-cylinder engine described in chapters 5 and 6, measurements about soot and particulate number production were carried out at TU Berlin by Mrs. Maike Gern.

The particle size distribution was determined with the mobility spectrometer DMS500 from Cambustion, which enables the analysis of particles with a specific diameter from 5 to 1000 nm. This method is based on the electrometer principle in which the particle size is classified on the basis of electrical mobility. Water injection had a visible influence on the particulate number concentration (PNC) depending on the injection concept and strategy. For example, with increasing W/F ratio, the PNC increased for DWI as well IWI due to the decrease of combustion stability (see for detail the paper [21]). Besides, a shift of the start of water injection (SOI_W) influenced the PNC. The dependency of the particle size distribution with respect to SOI_W variation are reported in figure 7.1 [21], for 30 % W/F, considering OP2. The particle number concentration (N) is represented logarithmically over the different size classes (Dp) and is given in dN/dlogDp [21]. A shift from an early to late SOI_W (210 to

90 °CA b.FTDC) raised the PNC consistently, as shown in figure 7.1. Probably the soot oxidation is prevented by the lower combustion temperature due to water injection. For further investigation CFD simulations were performed using the model of the single-cylinder engine described in chapters 5 and 6. As shown in the next section, the simulations provided an indirect explanation for the particle number increase by late SOI_W.

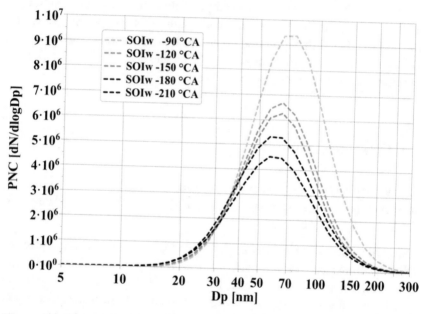

Figure 7.1: Measurements of particle size distribution in accumulation mode depending on the start of water injection for OP2 [21].

7.2 Simulation results

The water injection timing plays a fundamental role in the soot formation, as shown from the experimental data. Therefore, the influences of the water injection timing were investigated with 3D-CFD simulations for OP2 and DWI, as reported in table 7.1.

Table 7.1: 3D-CFD simulation for DWI and SOWI variation

Test	λ	Water injection pressure [bar]	W/F ratio [%]	SOWI [°CA b.FTDC]
DWI 1	1	200	30	120
DWI 2	1	200	30	210

Regarding figure 7.2, the flow field of the two DWI strategies is shown at 4 °CA a.FTDC.

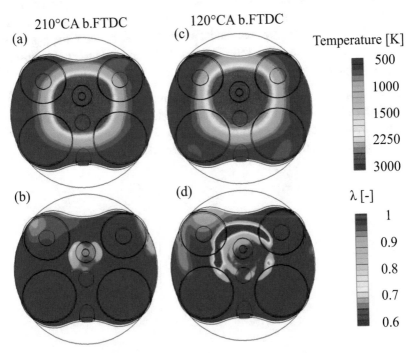

Figure 7.2: Temperature and air to fuel ratio distribution at 4 °CA a.FTDC, for two different water injection timings, respectively at 210 °CA b.FTDC and 120 °CA b.FTDC, at OP2 [21].

Particularly, with reference to figure 7.2.a and figure 7.2.b an early water injection timing is considered. According to experimental data, this should correspond to the case with moderate soot production. On the other side, pictures (c) and (d) represent the case with late water injection, which means a SOI_W regulated almost to the end of the compression stroke. At 120 °CA b.FTDC water injection, the effects of water vapour on mitigating the temperature of the combustion phase is higher (c) to the early injection (a). At the ignition point, the water vapour content of the mixture for the early injection is higher than the one of the late injection. Thus a stronger post evaporation event happens during the combustion process for the case with late water injection, bringing to a higher local decrease of the combustion temperature. The presence of water droplets before the combustion starts, for the late case, and their not homogeneous evaporation, causes the charge mixture being locally more reach compared to the case of early injection. As shown in figure 7.2, the late injection (d) features a local fuel accumulation in the region where the gas in the combustion chamber reaches 1500 K, typical temperature and fuel concentration for the grow of soot particles.

The solution towards optimized water injection strategies must investigate the in-cylinder charge motion and the mixture evolution during combustion. Optimized water injection targeting could help to split water over larger cylinder area, thus reducing the probability of soot formation.

8 Water Injection for Exhaust Gas Temperature Control

This chapter deals with the investigation of OP3, corresponding to rated power (4500 rpm 20, bar IMEP), generally critical for exhaust gas temperature management. Since executing this load point at the test bench was very critical for problems related to the cooling of the cylinder head, the entire analysis was conducted by exploiting the potential of the *Virtual test bench*.

As previously mentioned the choice of 30 % W/F ratio, is a good compromise between water beneficial effects and its consumption. Moreover, for the rated power point 30 % W/F ratio is the minimum amount of water that allows the engine running at $\lambda = 1$ for all the next tested injection strategies, by keeping exhaust gas temperatures before turbine below the limit (this target corresponds to the limit set by fuel enrichment case).

Especially for IWI cases, if water evaporation increases engine volumetric efficiency, boost pressure was regulated to keep the engine at $\lambda = 1$. The following analysis consists in 3D-CFD simulations, where the entire flow field of the intake and exhaust system together with the combustion event is simulated, considering 70 % initial relative humidity of the ambient air, 200 bar direct gasoline injection at 320°CA b.FTDC. The valve timing is characterised by a delay of the exhaust valve closure of 30°CA, while the intake valve opening is anticipated of 10 °CA both to the cam rest position (see as reference table 5.2). In this configuration, at the rated power point, the engine operates with 2 % residual gas and 5 mg/cyc back-flow at IVC. The overall mass in the cylinder at ignition point is 714 mg/cyc.

As mentioned in chapter 5.1, the single-cylinder engine at the test bench was equipped with indirect water injector (single, low-pressure injector, 8 holes, water-optimised) placed in the intake manifold (-IM) in position (a), concerning figure 8.1 and with a direct water injector located laterally (-L) in position (b), with no-optimised water spray targeting (single high-pressure injector, 5 holes, conventional gasoline solenoid valve).

© The Author(s), under exclusive license to
Springer Fachmedien Wiesbaden GmbH, part of Springer Nature 2021
A. Vacca, *Potential of Water Injection for Gasoline Engines by Means of a 3D-CFD Virtual Test Bench*, Wissenschaftliche Reihe Fahrzeugtechnik

In the modelled engine mesh was possible to add two new injector housings to the original test bench configuration, namely location (c), indirect water injector, valve selective (-VS), injecting directly on the valve seat and a centrally-mounted (-C) direct water injector in position (d) close to the gasoline one. The set-up of the simulation environment in QuickSim and the different virtual water injector position were discussed in chapter 5.2. The base engine calibration, without water injection and 4 °CA b.FTDC spark advance produced a regular combustion event (no knocking combustion). The analysis focused on T_3 reduction by running the engine at $\lambda = 1$ and keeping constant the combustion phasing.

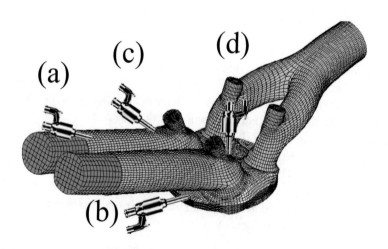

Figure 8.1: Injector configurations for IWI (a, c) and for DWI (b, d). Position (a) corresponds to IWI-IM (intake manifold) configuration, position (c) to IWI-VS (valve selective) one, position (b) as DWI-L (lateral) one and position (d) as DWI-C (central) one.

The potential of further advance the spark timing with water injection are investigated for this load point in chapter 8.1.1. The spark timing sweep was carried out just for IWI strategies since indirect water injection was found to be the optimal strategy for reducing T_3. The flow fields under water injection conditions and the mixture formation are studied for all the engine configurations mentioned above, for different injection timings and pressures. Residual gas and water of previous cycles are considered in the model together with their reciprocal influence to reproduce, for example, the water evaporation triggered by the higher temperature of residual gas. Since the CFD mesh is extended to the whole engine flow field, the water evaporation occurring in the exhaust runner is as well taken into account, especially for T_3 evaluation.

8.1 Indirect water injection strategies

Table 8.1 resumes the test matrix for the simulated IWI cases. In order to control exhaust gas temperature, the engine needs a fuel enrichment with $\lambda = 0.85$. These engine settings are consistent with the first test of table 8.1 and the maximum temperature in the exhaust runner obtained (930 °C) was used as the limit for the evaluation of the water injection case, with stoichiometric combustion.

The IWI tests were carried out at 30 % W/F ratio, with variable injection timing range from 390 °CA b.FTDC (early injection) to 244 °CA b.FTDC (late injection). The injection window was selected with the condition of end of water injection before IVC. Commercial indirect water injectors are nowadays able to reach an injection pressure of 10 bar. For the current engine analysis, 15 bar injection pressure was mostly used to enhance water evaporation and to reach a comparable evaporation rate, realized with the investigated DWI case at 100 bar. An additional simulation, with 7 bar injection pressure, was as well considered, in combination with later injection timing to enhance water evaporation.

With early IWI-IM in configuration (a), concerning figure 8.1, and 7 bar injection pressure, the amount of evaporated water mass reached 8 mg at ignition point, meaning an evaporation rate of 6-7 % (by already subtracting the initial water vapour content due to 70 % initial RH).

Table 8.1: Test matrix for indirect water injection cases at 30 % W/F ratio at rated power (OP3).

Test	λ [-]	Water injection pressure [bar]	Water injector position	SOI_{water} [°CA]
NoW $\lambda = 0.85$	0.85	-	-	-
NoW $\lambda = 1$	1	-	-	-
IWI-IM Early	1	15	Int. runner (a)	-390
IWI-IM Late	1	15	Int. runner (a)	-244
IWI-IM Late 7	1	7	Int. runner (a)	-244
IWI-VS Early	1	15	Valve select. (c)	-390
IWI-VS Late	1	15	Valve select. (c)	-244

With reference to figure 8.2, the in-cylinder water mass balance for the simulations of table 8.1 is presented. The curves report the " fresh " water vapour mass in the cylinder, meaning before being crossed by the flame front. The cases with no water injection (black and dark grey lines, respectively for $\lambda = 0.85$ and $\lambda = 1$) show the contribution of initial RH of air (up to 7 mg/cyc water vapour at ignition point arose from initial RH of air). Following these last lines, after FTDC, almost all the water vapour is crossed by the flame front, and the fresh water vapour mass disappeared. Indeed the fresh water vapour mass is transformed into " burned water vapour " by the flame front passing through it. This newly implemented feature in QuickSim was particularly helpful to analyse water effects during the different engine processes, because it allowed the decoupling of water taking part to combustion event (burned water vapour), the water influencing the laminar flame speed, the water vaporising during the combustion process and the one not crossed by the flame front at all. With this distinction (see chapter 3.3) one can highlight two different events for the water evaporation, one accomplishing during the compression stroke up to FTDC, and one realising during and particularly after combustion.

Early water injection strategies present lower evaporation rate (circle-markers in 8.2). Delayed water injection timings compensate the lower evaporation rate caused by adopting for example lower injection pressure (square-markers in 8.2). Particularly effective is the influence of water accumulated in the previous engine cycle. Considering the simulation *IWI-IM Late*, (late injection and configuration (a) with respect to figure 8.1), the furthest injection position with respect to cylinder shows a steeper increasing gradient compared to the other cases (triangle-markers line), and the rise in relation to the simulation with no water injection, starts even before the respective water injection. The reason is that a fraction of the injected water of the previous cycle is still present in the intake manifold during the analysed cycle and the water evaporation starts even before injection.

The in-cylinder temperature reduction at ignition point is comparable for all IWI strategies, namely 20 °C decrease to the case at $\lambda = 1$ and no water injection and comparable with the fuel enrichment case.

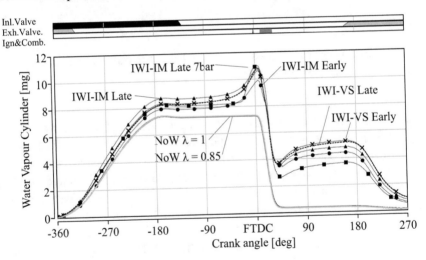

Figure 8.2: Fresh water evaporated mass during pre and post evaporation event with different IWI strategies, for rated power (OP3).

Figure 8.3 reports the liquid water mass detected in the liner (by isolating the liner cells in the mesh), particularly interesting during the compression phase. The diagram shows an accumulation of 2 mg of liquid water in the cylinder liner region, almost for all the tested strategies. The worse case was represented by *IWI-VS Early*, where the liquid water accumulation starts already at 300 °CA b.FTDC.

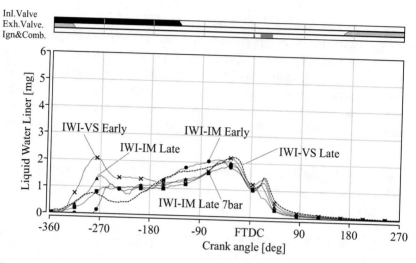

Figure 8.3: Water liquid mass detected at liner with different IWI strategies, for rated power (OP3).

Concerning the water evaporation trend again, when valve selective injector position is considered (x-markers and dashed black line in figure 8.2) a sufficient evaporation rate is realised at ignition point, but a higher post evaporation event takes as well place as visible in figure 8.2 between 90°CA and 180°CA a.FTDC. The post evaporation, realised in the final combustion phase, becomes particularly effective for T_3 reduction as pictured in figure 8.4. All IWI strategies with 30 % W/F ratio resulted in a comparable T_3 decrease, compared to fuel enrichment case (dark grey curve in figure 8.4), leading in most cases to further improvement. Among these, the best result is achieved by late injection right away above the intake port (*IWI-VS Late*).

With reference back to the analysis provided in figure 8.2, the post evaporation of *IWI-VS Late* case, which presented the most important rate among all simulations, it produced the best result in terms of T_3 decrease, as it can be seen in figure 8.4. On top of that, it must bear in mind that all IWI simulations featured a stoichiometric mixture and thus the gain in term of T_3 reduction to no water injection case and $\lambda = 1$ was around 100 K. The small diagram in figure 8.4, resumes the exhaust gas temperatures detected for every IWI strategies at 220

°CA a.FTDC. Furthermore, *IWI-VS Late* configuration was the one which also showed the more significant temperature reduction in the unburned gas zones (the maximal unburned gas temperature decreased from 530 °C to 470 °C considering the case with no water injection and $\lambda = 1$), thus enhancing as well Knock resistance and opening the possibility to increase the spark advance.

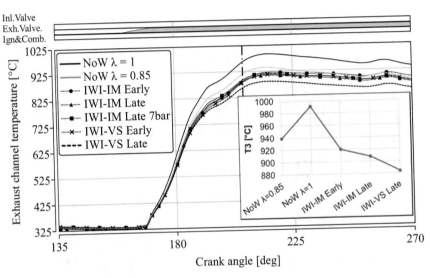

Figure 8.4: Exhaust gas temperature before turbine (T_3), with different IWI strategies, for rated power (OP3).

8.1.1 Ignition point sweep at rated power

Considering the self-ignition model described in chapter 3.4 and used as well in chapter 6, it was possible to optimise the spark advance of the rated power load point. Nevertheless, this analysis has to be considered just as first assessment since no measurements were available for the rated power point concerning Knock occurrences. Therefore, the mass in self-ignition condition considered as knocking combustion threshold was estimated based on other engine data. About figure 8.5, the fuel enrichment case is compared with the stoichiometric case and with the $\lambda = 1$ case and IWI, at constant spark advance (4 °CA b.FTDC). It was possible then to advance the ignition point up to 7°CA b.FTDC (without hitting the Knock limit) and to enhance as well the engine indicated efficiency to the case at $\lambda = 1$ (which already realises a considerable fuel saving and thus efficiency increase to the fuel enrichment case).

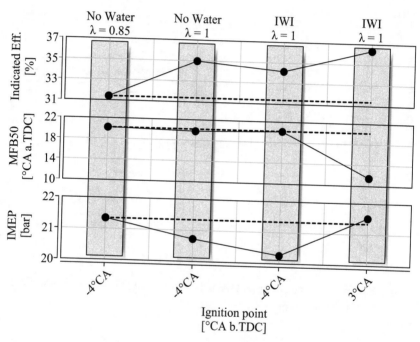

Figure 8.5: Indicated efficiency, MFB50 and IMEP by varying the ignition point for IWI strategies at rated power (OP3).

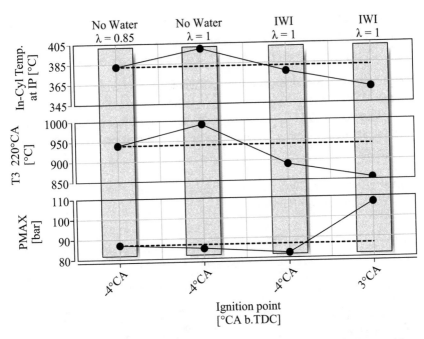

Figure 8.6: In-cylinder temperature, exhaust gas temperature before turbine (T_3) and PMAX varying the ignition point at rated power (OP3).

8.2 Direct water injection strategies

Table 8.2 presents the simulation matrix for DWI strategies for OP3 (4500 rpm, 20 bar IMEP). These simulations consider just stoichiometric mixture conditions (no fuel enrichment needed for component protection), injecting water at 100 bar with different injector positions in the combustion chamber and in combination with the variation of the injection timing. The tests were all carried out at 30 % W/F ratio, as already done for IWI strategies.

The investigated engine test bench configuration was at first corresponding to the original engine layout, with lateral water injector, which means layout (b) of figure 8.1, abbreviated as *DWI-L* (lateral) and considering as water injection timing 120 °CA b.FTDC. The experiments corresponding to this configuration were used to calibrate the respective simulation (reference case).

Being the calibration of simulation realized for the reference case, a central injector (C.) was just virtually tested. In addition, an injection timing variation was realized from 330 °CA b.FTDC to 20 °CA a.FTDC, meaning that a post-injection strategy, injecting in the middle of the combustion event, was as well investigated.

Table 8.2: Test matrix for direct water injection cases at 30 % W/F ratio, at rated power (OP3).

Test	λ [-]	Water injection pressure [bar]	Water injector position	SOI_{water} [°CA]
DWI-L Late	1	100	Lateral (b)	-120
DWI-C Late	1	100	Central (d)	-120
DWI-C Early	1	100	Central (d)	-330
DWI-C Post	1	100	Central (d)	20

The water mass balance analysis is examined for DWI simulations in figure 8.7. Here a comparison between lateral (*DWI-L Late*) and central (*DWI-C Late*) water injection can be done for late injections (x-markers versus circle-markers curves of figure 8.7). Once injecting water laterally, the evaporation rate at the ignition point reached 16 % compared to the overall water injected quantity, while the evaporation rate improves up to 25% when the central injector was used.

Such a difference in the amount of vaporized water at the ignition point generated a decrease of 6 °C of in-cylinder temperature before the combustion starts, by observing figure 8.8.

At the end of the combustion process, *DWI-L Late* showed a higher presence of fresh water vapour mass (figure 8.7), to *DWI-C Late*. Moreover, these DWI strategies registered the absence of an increasing gradient for water mass evap-oration after combustion, as it was instead featured for the IWI cases reported in figure 8.2 (see chapter 8.1). As a result, very low post-evaporation realized for DWI strategies, to IWI ones.

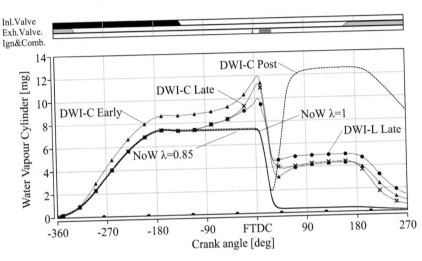

Figure 8.7: Fresh water evaporated mass during pre and post evaporation event with different DWI strategies, at rated power (OP3).

Furthermore, the higher presence of fresh water vapour at the end of combustion for *DWI-L Late* case (circle-markers line) proofed that with lateral injector less water vapour transited across the flame front with respect to the central injector (x-markers line in figure 8.7), generating the desired mitigation effect of water injection. For this reason, the central injector was expected to be more effective for T_3 reduction to the lateral one, but the DWI strategies were as well assumed to be less effective for the reduction of T_3, with respect to IWI, because of the lower post-evaporation event.

Using the central injector again, but with early injection timing (triangle- markers line in figure 8.7), the evaporation rate at ignition point reached 30 %, lowering the in-cylinder temperature by 12 °C to *DWI-L Late*, before the combustion starts, as it can be observed in figure 8.8. The time available for the vaporization, due to the early injection timing, made the evaporation effect distributing over more mass, exploiting more the water cooling power.

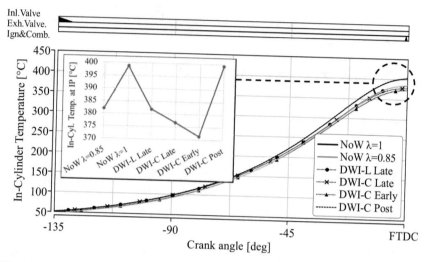

Figure 8.8: In-cylinder temperature from intake valve closure up to ignition point, for different DWI strategies, at rated power (OP3).

On the other side, following the respective DWI strategies in figure 8.9 it can be seen that *DWI-C Early* generated the higher accumulation of water droplets at the liner since water injection realized at high pressure in a low-density flow field (330 °CA b.FTDC corresponds to the first phase of the compression stroke). This result would probably cause more significant issues if considering oil dilution. By comparing liquid mass at liner for IWI and DWI strategies respectively of figure 8.3 and of figure 8.9, DWI strategies seemed more critical for liquid water presence at liner, bringing 50 % more water accumulation (6 mg against 4 mg in the worst case).

Concerning figure 8.7 and particularly the post-injection case (dashed black line), the post-evaporation of fresh water reached the highest gradient in comparison to other strategies. Here locally, the conditions for a flash boiling of injected water were achieved due to very high temperatures at the combustion end. This phenomenon was, unfortunately, a local event, happening as well in a narrow time window, as shown in figure 8.10, where the 3D plots compare in-cylinder temperature for *DWI-C Early* strategy with *DWI-C Post* one.

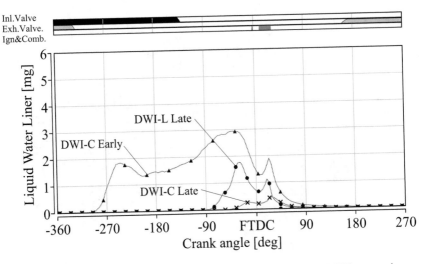

Figure 8.9: Water liquid mass detected at liner for different DWI strategies, at rated power (OP3).

The cylinder head on the right of figure 8.10, reveals indeed, a local drop of temperatures in the area where water is injected, due to instantaneous vaporization. Moreover, the post-injection had not the possibility to decrease in-cylinder temperature before the start of combustion (see again figure 8.8) and the global temperature reduction, due to water injection, accomplished later.

The results in terms of T_3 reduction are resumed in figure 8.11. The *DWI-L Late* temperature decrease (circle-markers line) is comparable with the fuel enrichment case (grey line) and at the limit of the engine bearable T_3 peak temperature. The first improvement of 20 °C T_3 reduction accomplished by moving the injector to the central position (even without optimizing the spray targeting). The higher evaporation rate realized by *DWI-C Early* contributed to the greater T_3 decrease (triangle-markers line of figure 8.11). The water post-injection, even featuring the highest post evaporation event, was not able to properly affect the temperature chain reduction, losing part of water cooling effect before ignition point and during the first phase of combustion.

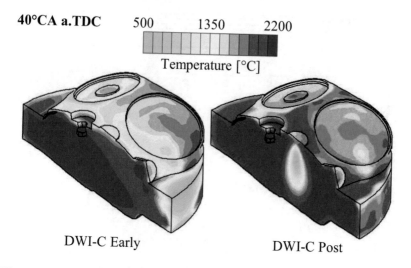

Figure 8.10: In-cylinder temperature for *DWI-C Early* and *DWI-C Post*, with 30 % W/F ratio at 40 °CA a.FTDC, for rated power (OP3).

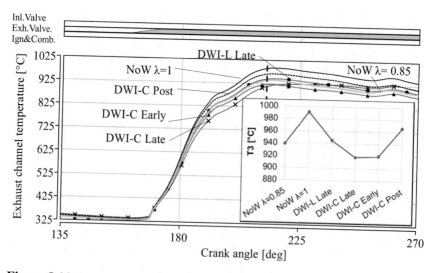

Figure 8.11: Exhaust gas temperature before turbine (T_3), for different DWI strategies, at rated power (OP3).

8.3 Comparison between indirect and direct water injection

Through IWI and DWI strategies, there is the possibility to control differently the parameters that regulate water effects on engine processes, such as water distribution and time available for the evaporation. For IWI, water distribution is more sensible to injection timing rather than to injection pressure, while water evaporation for DWI cases is strictly related to injector position.

Figure 8.12 points out evaporated water mass concentration at ignition point and in the cylinder, for three different DWI strategies comparing them with the most effective IWI approach (*IWI-VS Late*). The lateral direct injector, placed below inlet valves, produces an accumulation of water droplets between the exhaust valves. Here the temperature field supported their evaporation, and water vapour converged in a narrow area surrounding the exhaust valves, as it can be seen in figure 8.12.a. The local accumulation of the water prevented the water evaporation itself, and indeed the *DWI-L Late* strategy had the worst results in terms of evaporation rate at ignition point and consequently the smaller T_3 reduction, among the configurations shown in figure 8.12. An improvement step accomplished by moving the direct injector to a central position, allowing a more uniform mixture formation, keeping the injection timing at 120 °CA b.FTDC (figure 8.12.b). By advancing injection timing to 330 °CA b.FTDC (figure 8.12.c), at the same injection pressure of 100 bar in relation to the other DWI strategies, a huge enhancement of water distribution was reached. A gradual evaporation of water took place, as reported in figure 8.7 (chapter 8.2), and the time available for water droplets distribution and their evaporation, produced a uniform water vapour mass concentration in the whole cylinder before combustion. This higher evaporation rate and better spatial distribution led to control in-cylinder temperature during combustion (roughly 50 K lower peak temperature to the no water case with fuel enrichment) allowing the engine being more resistant to Knock, decreasing T_3 and NOx engine out formation. Good water distribution was as well realized by IWI when the injector was placed directly on the intake valve with late injection timing (figure 8.12.d).

Figure 8.13 presents the results of the 3D-CFD virtual test bench for different IWI and DWI configurations.

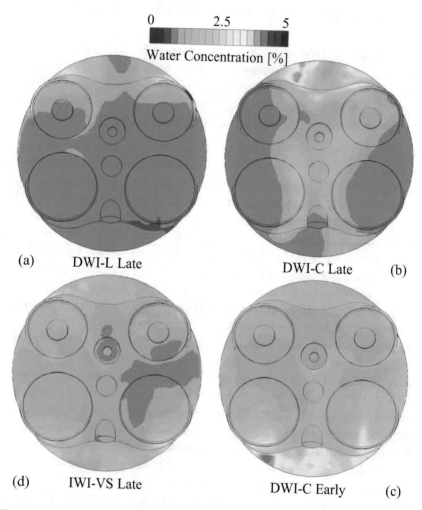

Figure 8.12: Water vapour concentration at ignition point, comparing IWI
and DWI strategies and injector positions, at rated power (OP3).

Considering the x-axis, the first two simulations correspond to the cases without
water injection, respectively, with and without fuel enrichment. IWI strategies
are then reported, together with DWI cases towards the end of the axis. For

each simulation is reported liquid water mass at liner (square-markers lines) considering 45 °CA b.FTDC, namely the most critical point for oil dilution during the compression stroke. The circle-markers lines represent liquid water mass at 225 °CA a.FTDC, going towards the exhaust gas line. This liquid water mass is as well a primary parameter for evaluating the water injection strategy because the liquid water could reduce the *Three-way catalyst* (TWC) efficiency. Immediately below, two curves report the evaporation rate of the overall injected water before the start of combustion (star-markers lines) and in-cylinder temperature conditions at ignition point (x-markers lines). The triangle-markers lines, instead, show exhaust gas temperature before the turbine at the position of T_3 sensor in the engine mesh.

The water mass accumulation at the liner was more critical for the *DWI-C Early* case. As a result, by injecting at 100 bar pressure, directly into the cylinder, at 330 °CA b.FTDC, caused water impinging the wall severely and collecting up to 3 mg liquid water. The *DWI-C Late* simulation instead, showed the lower water accumulation tendency at the liner. *DWI-L Late* was a compromise among DWI strategies in terms of water accumulation at liner. On the other hand, all IWI strategies brought gathering 2 mg of liquid water at the liner, regardless of injection pressure or timing. *IWI-IM Late 7* concept led to the highest amount of liquid mass in the exhaust manifold (the higher water stagnation in the intake manifold let water droplets going towards exhaust manifold during valve overlap), while generally, DWI was preferable to IWI strategies in order to preserve the TWC.

Considering the effects at the ignition point, the fuel enrichment case, without water injection, decreased the in-cylinder temperature up to 20 K to the case at $\lambda = 1$. Comparable results were achieved by all IWI strategies, while a further reduction of 8 K, in relation to the fuel enrichment case, was realized by the *DWI-C Early* configuration. No effects on the in-cylinder temperature from the post-injection were expected neither observed. Water evaporation rate at ignition point for IWI strategies shows that late injection timings are advisable and even compensating the effect of lower injection pressures. Evaporated water mass depends as well on the water mass still present in the intake manifold from the previous cycle; thus 7 bar injection pressure could feature a higher evaporation rate with respect to 15 bar.

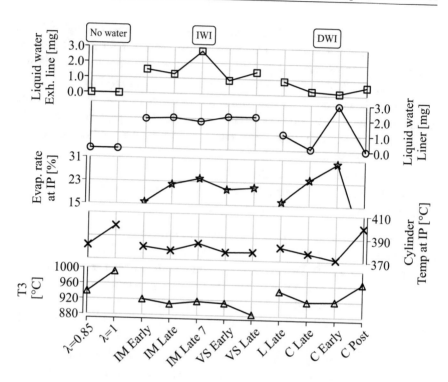

Figure 8.13: Resume of the main engine output parameters with different IWI and DWI configurations, at rated power (OP3).

Moving the direct lateral water injector to a central position and with early timing brings a significant increase of evaporated water mass. The trend of T_3 reflects almost entirely the one of the in-cylinder temperatures. The fuel enrichment cooling power for T_3 control was in the range of 50°C temperature decrease. Water injection cases could keep exhaust gas temperature before the turbine below such a limit with $\lambda = 1$ and even improve the cooling down performance to the fuel enrichment case. IWI strategies result particularly effective for T_3 reduction. The *IWI-VS Late* case realized the maximum T_3 reduction, lowering exhaust gas temperature of a further 40 K, with respect to the best DWI strategy. The post-injection strategy is the only one that did not reach acceptable result for T_3 reduction.

9 Conclusion and Outlook

Among others, water injection is one of the most promising technologies to improve the engine combustion efficiency, by reducing Knock occurrences, controlling exhaust gas temperature and allowing the engine operating at $\lambda = 1$ over the whole engine map.

Water injection implementation accomplishes through different engine layouts and injection strategies. Besides, the results of injecting water are strictly dependent on the base engine features and load point. Therefore, the engine virtual development becomes essential to study many engine configurations. Considering the different numerical approaches, 3D-CFD simulations are a powerful and essential tool especially for an in-depth analysis of water injection effects in gasoline engines. The *Virtual Test Bench*, thanks to its ability to cope with new operating parameters and hardware solutions, gives insights and results even more significant for the implementation of water injection in series engines.

In this dissertation, indirect (IWI) and direct water injection (DWI) strategies were evaluated to increase the engine Knock limit and reduce exhaust gas temperature before turbine (T_3) for three different operating points.

9.1 Optical measurements and 3D-CFD injection simulations

In parallel with the implementation of water thermodynamics in QuickSim, a new methodology for coupling optical measurements and 3D-CFD simulations was developed with the intent of increasing the reliability of the injection simulations and with the additional objective of investigating the spray characteristics, together with the virtual engine development. This approach led to enhance the settings of the break-up model and particularly the calibration of the droplets initialization domain, which in turn meant the calibration of the spray pattern and penetration within the combustion chamber [50]. The

A. Vacca, *Potential of Water Injection for Gasoline Engines by Means of a 3D-CFD Virtual Test Bench*, Wissenschaftliche Reihe Fahrzeugtechnik

developed procedure represents a global approach for the spray geometry analysis and therefore for the design of new injectors, accomplished by taking into account together mixture formation (in a real engine flow field), influence on the combustion process and Knock onset, but as well wall impingement issues.

9.2 Effects of indirect and direct water injection strategies

The numerical implementation of water injection in QuickSim and the spray investigation were the fundamentals of the further full engine analysis.

The adoption of DWI instead of IWI, for OP1 (2000 rpm, 20 bar IMEP) brought many advantages. This load point was characterized by a rather delayed ignition point, because of the high Knock sensitivity with the base calibration. Especially intake manifold water injection was not able to provide a sufficient charge cooling effect before ignition point and therefore, its influence in reducing Knock tendency was lower. Independently of the applied injection strategy, by keeping 30 % water to fuel ratio, IWI was not effective for improving the centre of combustion (MFB50) at this load point. On the other side, almost all DWI tested strategies led to improve the combustion phasing with two separate mechanisms. First, the charge cooling effect was enhanced by injecting water directly into the cylinder, with a favourable flow field for water evaporation, together with smaller droplets and higher jet velocities. Secondly, DWI could be used to improve the charge motion within the cylinder. In this sense, developing a specific spray angle and spray pattern the engine Tumble could be fostered and therefore, the turbulent kinetic energy boosted. This second effect improved mixture formation and triggered faster combustion, making the flame front running over the End-gas region sooner and thus enhancing Knock resistance. DWI could be even employed in some case for increasing in-cylinder pressure, without changing the spark advance.

In detail, at OP1, the combination of DWI at 200 bar with water-developed spray targeting (IVK T), it allowed to reduce the temperature of the charge, to improve mixture formation, due to higher turbulence, and the spark advance could be set earlier, since the enhanced Knock resistance. As summarized in figure 9.1, the MFB50 moved from 21 to 12 °CA a.FTDC, shortening combustion duration and leading to an increase in the engine indicated efficiency up to 12 %. The load point was always reached with the same injected fuel, keeping stoichiometric mixture, which means that more efficient combustion results in higher torque [49].

Since the interaction between injected water and charge motion, depends on injection pressure, timing and targeting, it is now clear that at first glance, it is not possible to assess whether a higher water injection pressure would achieve the best results. For this reason, the water injection strategy has to be investigated with respect to the load point and the injector layout, individually for each engine configurations.

Considering OP2 (2500 rpm, 15 bar IMEP), with reference again to figure 9.1, the injection of water, both indirectly or in the combustion chamber, produced no advantages in terms of combustion phasing and therefore for fuel consumption. At OP2 the base engine was already calibrated with an earlier spark advance, and no improvement margins were found. On the other side, the emission benefits were clear, considering NOx formation. For the soot production instead, it was found that the start of water injection played a key role.

IWI achieved the most important result for reducing exhaust gas temperature before the turbine. Especially for OP3, the rated power point of the engine (4500 rpm, 20 bar IMEP), IWI showed more important potential for T_3 reduction, to DWI. This result was accomplished by moving the injector from the intake manifold directly on the intake port. No evident influences among 5 and 15 bar injection pressure were highlighted. Late injection timing compensated for lower injection pressure, in terms of T_3 reduction. For IWI, water distribution is more sensible to injection timing rather than to injection pressure, while water evaporation for DWI cases is strictly related to injector position. Overall, both IWI and DWI allowed avoiding the fuel enrichment completely with 30 % water to fuel ratio. The rated power operating point could be optimized, lowering fuel consumption up to 12 % with respect to the base calibration, as shown in figure 9.1. The possible reduction of T_3 with water injection and stoichiometric combustion was 85 K compared to the equivalent case without water injection. Besides, by comparing the fuel enrichment case with the best case of water injection and stoichiometric combustion, a reduction of T_3 of 40 K was further accomplished, compared to the rich combustion case [48].

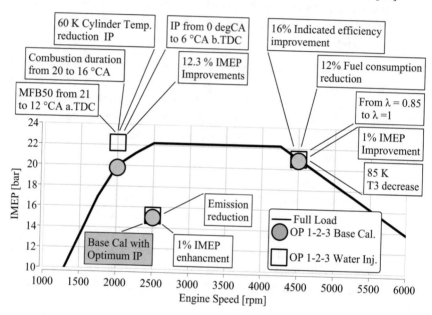

Figure 9.1: Single-cylinder engine map, with base calibration and with enhanced performances realised through water injection for 3 operating points.

The water effects are strongly dependent on engine configuration and particularly on load point as resumed in table 9.1. IWI strategies are not less effective compared with DWI, at least for T_3 reduction. By optimizing the water injection strategy, IWI can realize the same exhaust gas temperature decrease as DWI and even improve it, thanks to the exploitation of the post evaporation process (see chapter 8). IWI configurations show potentially less wall impingement and thus a lower oil dilution effect. As a drawback, IWI increases the possibility of intake manifold condensation and uneven water distribution over the cylinders. On the other side, DWI strategies enable to control better in-cylinder temperature enhancing the Knock resistance and improving the engine indicated efficiency as it was realized for OP1. The design of a direct water-optimized injector targeting plays a key role in the engine response to water injection. More extensive spray targeting allowed increasing the air entrainment, thus promoting water evaporation, but an evaluation of the liquid amount impinging the wall always has to be encompassed simultaneously.

Table 9.1: Resume of the results about the main engine parameters for the investigated load points with IWI and DWI strategies.

	OP1 2000 rpm 20 bar IMEP		OP2 2500 rpm 15 bar IMEP		OP3 4500 rpm 20 bar IMEP	
	IWI	DWI	IWI	DWI	IWI	DWI
T_3 reduction	++	++	++	++	++	+
Wall impingement	++	-	++	-	++	-
Knock resistance	/	++	/	/	+	++

9.3 Effects of Miller valve profile

Finally, the experimentally validated engine model was used to study the possible rise of ε, in combination with the Miller strategy and DWI for OP1. The Miller lift was generated, keeping the same valve acceleration and considering different maximum lifts. The reduced intake air mass, combined with the water cooling effect led to adopt higher ε (12 : 1 instead of 10.75 : 1), without overcoming the Knock threshold and adding to the already enhanced indicated efficiency, a further improvement margin due to the longer expansion effect (2 % additional efficiency gain on top of 12 % overall advantage of DWI with IVK T targeting at 200 bar).

The analysis led in this dissertation was marked out as *3D-CFD Virtual Test Bench* because of the huge amount of configurations tested through simulations and for the similarity to the experimental methodology usually adopted for the investigation of the engine thermodynamics at the test bench. The set-up of the *Virtual Test Bench* was exploited to evaluate earlier spark advance to Knock onset, together with a higher compression ratio in combination with Miller valve strategy and DWI. The results showed the essential contribution of fast 3D-CFD simulations for increasing engine efficiency through water injection and proposing a new approach for the future power-train engine development process.

9.4 Outlooks and further possibilities

As possible further steps, the investigation of water to fuel emulsion should be done. The emulsion adoption imposes new challenges for the adjustment of water injection strategy to load variations since the engine regulation should provide the rail always with the desired W/F ratio.

The investigation of the water droplet breakup through *Phase Doppler Anemometry (PDA)* for different injectors can bring new knowledge on water evaporation and the possibility to enhance it by reducing wall impingement and thus oil dilution.

Among the different DWI strategies, one further possibility to enhance water evaporation is the adoption of a multi-point water injection strategy.

Water injection can also be investigated as a mean to reduce eventually EGR rate in those engine applications which require such regulation for load control or emission reduction.

The effect of water addition on the combustion and emission formation has continued to be investigated for different RON sensitive fuels. On top of that, experiments and simulations of synthetic fuels will add new knowledge to the different means of optimizing the normal combustion process.

Bibliography

[1] World weather and climate information (https://public.wmo.int/) - 2019.

[2] Market penetration analysis of electric vehicles in the german passenger car market towards 2030. *International Journal of Hydrogen Energy*, 38(13):5201 – 5208, 2013.

[3] *Grundlagen Verbrennungsmotoren*. Springer Fachmedien Wiesbaden, 2014.

[4] J. Anderson. *Computational Fluid Dynamics*. McGraw-Hill Education - Europe, 1995.

[5] A. Babajimopoulos, D. Assanis, D. Flowers, and et al. fully coupled computational fluid dynamics and multi-zone model with detailed chemical kinetics for the simulation of premixed charge compression ignition engines. In *International Journal of Engine Research*, 2005.

[6] M. Baratta, A. Catania, E. Spessa, L. Herrmann, and K. Roessler. Multi-dimensional modeling of direct natural-gas injection and mixture formation in a stratified-charge si engine with centrally mounted injector, apr 2008.

[7] M. Battistoni, C. Grimaldi, V. Cruccolini, G. Discepoli, and M. De Cesare. Assessment of port water injection strategies to control knock in a gdi engine through multi-cycle cfd simulations. In *13th International Conference on Engines & Vehicles*. SAE International, Sep 2017.

[8] J. Blazek. *Computational Fluid Dynamics: Principles and Applications*. Elsevier Science, 2005.

[9] C. Buchal, H.-D. Karl, and H.-W. Sinn. Kohlemotoren, windmotoren und dieselmotoren: Was zeigt die co2-bilanz?

[10] CD adapco. *Methodology - STAR-CD Version 4.20*, 2013.

[11] M. Chiodi. *An Innovative 3D-CFD-Approach towards Virtual Development of Internal Combustion Engines.* PhD thesis, University of Stuttgart, 2010.

[12] F. Cupo. *Modeling of Real Fuels and Knock Occurrence for an Effective 3D-CFD Virtual Engine Development.* PhD thesis, University of Stuttgart, to be published in 2020.

[13] F. Cupo, M. Chiodi, M. Bargende, D. Koch, and D. Winckelhaus. Virtual Investigation of Real Fuels by Means of 3D-CFD Engine Simulations. In *SAE Technical Paper Series*, number 2019-24-0090, 2019.

[14] V. De Bellis, F. Bozza, L. Teodosio, and G. Valentino. Experimental and numerical study of the water injection to improve the fuel economy of a small size turbocharged si engine, mar 2017.

[15] B. Durst, G. Unterweger, S. Rubbert, A. Witt, and M. Böhm. Thermodynamic effects of water injection on otto cycle engines with different water injection systems. In *15th Conference "The Working Process of the Internal Combustion Engine", Graz*, Sept 2015.

[16] H. Eichlseder, M. Klüting, and W. Piock. *Grundlagen und Technologien des Ottomotors.* Springer-Verlag KG, 2008.

[17] B. Franzke, T. Voßhall, P. Adomeit, and A. Müller. Wassereinspritzung zur erfüllung zukünftiger rde-anforderungen für turbo-ottomotoren. *MTZ - Motortechnische Zeitschrift*, 80:32–41, 03 2019.

[18] M. Gern, G. Kauf, A. Vacca, T. Franken, and A. Kulzer. Ganzheitliche methode zur bewertung der wassereinspritzung im ottomotor. *MTZ - Motortechnische Zeitschrift*, 80:124–129, 07 2019.

[19] M. Gern, G. Kauf, A. Vacca, T. Franken, and A. Kulzer. Integrated method for the evaluation of water injection in gasoline engines. *MTZ worldwide*, 80:114–119, 07 2019.

[20] M. Gern, M. Kauf, and R. Baar. Experimental investigation of water injection and strategy concepts for gasoline engines. *iMechE – Fuel Systems, London*, page 105–115, 2018.

[21] M. S. Gern, A. Vacca, and M. Bargende. Experimental analysis of the influence of water injection strategies on DISI engine particle emissions. In *SAE Technical Paper Series*, number 2019-24-0101, 2019.

[22] R. Golloch. *Downsizing bei Verbrennungsmotoren: Ein wirkungsvolles Konzept zur Kraftstoffverbrauchssenkung (VDI-Buch) (German Edition)*. Springer, 2005.

[23] C. Heinrich, H. Dörksen, A. Esch, and K. Krämer. Gasoline water direct injection (gwdi) as a key feature for future gasoline engines. pages 322–337, 12 2018.

[24] I. Hermann, C. Glahn, M. Paroll, and W. Gumprich. Water injection for gasoline engines, potentials and challenges. In J. Liebl, C. Beidl, and W. Maus, editors, *Internationaler Motorenkongress 2019*, pages 115–138, Wiesbaden, 2019. Springer Fachmedien Wiesbaden.

[25] F. Hoppe, M. Thewes, H. Baumgarten, and J. Dohmen. Water injection for gasoline engines: Potentials, challenges, and solutions. *International Journal of Engine Research*, 17(1):86–96, 2016.

[26] F. Hoppe, M. Thewes, J. Seibel, and A. Balazs. Evaluation of the potential of water injection for gasoline engines. *International Journal of Engine Research*, 10(5):2500–2512, 2017.

[27] D. Hung, D. Harrington, A. Gandhi, L. Markle, S. Parrish, J. Shakal, H. Sayar, S. Cummings, and J. Kramer. Gasoline fuel injector spray measurement and characterization–a new sae j2715 recommended practice. *SAE International Journal of Fuels and Lubricants*, 1, 04 2008.

[28] M. Hunger, T. Böcking, U. Walther, M. Günther, N. Freisinger, and G. Karl. Potential of direct water injection to reduce knocking and increase the efficiency of gasoline engines. pages 338–359, 12 2018.

[29] A. Kächele. *Turbocharger Integration into Multidimensional Engine Simulations to Enable Transient Load Cases*. PhD thesis, University of Stuttgart, 2019.

[30] V. Knop and S. Jay. Latest Developments in Gasoline Autoignition Modelling Applied to an Optical CAI Engine. In *Oil and Gas Science and Technology, Bd. 61*, 2006.

[31] Y. Liao and W. Roberts. Laminar Flame Speeds of Gasoline Surrogates Measured with the Flat Flame Method. In *Energy Fuels*, 2016.

[32] J. Livengood and P. Wu. Correlation of autoignition phenomena in internal combustion engines and rapid compression machines,. In *International Symposium on Combustion*, 1955.

[33] M. Mehl, W. Pitz, C. Westbrook, and H. Curran. Kinetic modeling of gasoline surrogate components and mixtures under engine conditions. In *Proceedings of the Combustion Institute*, 2011.

[34] Ömer L. Gülder. Correlations of laminar combustion data for alternative s.i. engine fuels. In *West Coast International Meeting and Exposition*. SAE International, aug 1984.

[35] G. P. Merker, C. Schwarz, and R. Teichmann, editors. *Combustion Engines Development - Mixture Formation, Combustion, Emissions and Simulation*. Springer-Verlag Berlin Heidelberg, 2012.

[36] F. Millo, S. Luisi, F. Borean, and A. Stroppiana. Numerical and experimental investigation on combustion characteristics of a spark ignition engine with an early intake valve closing load control. In *Fuel, Volume 121, Pages 298-310*, 2017.

[37] N. Neumann, N. Freisinger, G. Vent, and T. Seeger. Experimental investigation of miller cycle combustion technology with water injection. 19th Stuttgart International Symposium, Stuttgart, March 2019.

[38] N. Otsu. A threshold selection method from gray-level histograms. *IEEE Transactions on Systems, Man, and Cybernetics*, 9(1):62–66, Jan 1979.

[39] S. Paltrinieri, F. Mortellaro, N. Silvestri, L. Rolando, and et al. Water Injection Contribution to Enabling Stoichiometric Air-to-Fuel Ratio Operation at Rated Power Conditions of a High-Performance DISI Single Cylinder Engine. In *SAE Technical Paper Series*, number 2019-24-0173, 2019.

[40] T. Pauer, M. Frohnmaier, J. Walther, P. Schenk, A. Hettinger, and S. Kampmann. Optimization of gasoline engines by water injection. In *37th Wiener Motorenkongress*, 2016.

[41] C. Pera and V. Knop. Methodology to define gasoline surrogates dedicated to autoignition in engines. In *Elsevier Fuel Journal, Bd. 96*, 2012.

[42] C. Pfeifer. *Experimentelle Untersuchungen von Einflußfaktoren auf die Selbstzündung von gasförmigen und flüssigen Brennstofffreistrahlen.* PhD thesis, Karlsruhe Institute of Technology, 2010.

[43] Y. Ra and R. Reitz. A reduced chemical kinetic model for IC engine combustion simulations with primary reference fuels. In *Combustion and Flame, Bd. 155*, 2008.

[44] R. Reitz. *Atomization and other breakup regimes in a liquid jet.* PhD thesis, Princeton University, 1978.

[45] E. Rossi. *Improvements of 3D-CFD simulations of high-performance multi-hole injectors.* M. Sc. thesis, University of Stuttgart, 2019.

[46] J. Shi, P. Lopez Aguado, and N. Guerrassi. Understanding high-pressure injection primary breakup by using large eddy simulation and x-ray spray imaging. *MTZ - Motortechnische Zeitschrift*, 78:50–57, 05 2017.

[47] Y. Sun, M. Fischer, M. Bradford, A. Kotrba, and E. Randolph. Water recovery from gasoline engine exhaust for water injection. 04 2018.

[48] A. Vacca, M. Bargende, M. Chiodi, T. Franken, C. Netzer, M. S. Gern, M. Kauf, and A. C. Kulzer. Analysis of Water Injection Strategies to Exploit the Thermodynamic Effects of Water in Gasoline Engines by Means of a 3D-CFD Virtual Test Bench. In *SAE Technical Paper Series*, number 2019-24-0102, 2019.

[49] A. Vacca, F. Cupo, M. Chiodi, M. Bargende, O. Berkemeier, and M. Khosravi. The virtual engine development for enhancing the compression ratio of disi-engines combining water injection, turbulence increase and miller strategy. In *SAE Technical Paper Series, CO2 Reduction for Transportation Systems Conference, Turin*, number 20CO-0007, to be published in 2020.

[50] A. Vacca, S. Hummel, K. Müller, M. Reichenbacher, M. Chiodi, and M. Bargende. Virtual development of injector spray targeting by coupling 3D-CFD simulations with optical investigations. In *WCX SAE World Congress Experience*. SAE International, apr 2020.

[51] R. van Basshuysen, editor. *Ottomotor mit Direkteinspritzung und Direkteinblasung - Ottokraftstoffe, Erdgas, Methan, Wasserstoff*. Springer Fachmedien Wiesbaden, 2016.

[52] R. van Basshuysen and F. Schäfer, editors. *Handbuch Verbrennungsmotor*. Springer, 2014.

[53] B. VanDerWege, T. Lounsberry, and S. Hochgreb. Numerical modeling of fuel sprays in disi engines under early-injection operating conditions. In *SAE Technical Paper Series*, number 2000-01-0273, 2000.

[54] VDI-Gesellschaft Verfahrenstechnik und Chemieingenieurwesen, editor. *VDI-Wärmeatlas*. Springer, 2013.

[55] M. Wentsch. *Analysis of Injection Processes in an Innovative 3D-CFD Tool for the Simulation of Internal Combustion Engines*. PhD thesis, University of Stuttgart, 2018.

[56] M. Wentsch, M. Chiodi, and M. Bargende. Fuel injection analysis with a fast response 3d-cfd tool,. In *SAE Technical Paper Series*, number 2017-24-0103, 2017.

[57] M. Wentsch, M. Chiodi, M. Bargende, D. Wichelhaus, and J. B. von Fackh. 3D-CFD-Modeling of Multi-Component Liquid Gasoline for an improved Analysis of In-Cylinder Internal Engine Phenomena. Number 216-20135172 in JSAE Annual Congress (Spring), Yokohama (Japan), May 2013.

[58] J. Zammit, M. McGhee, P. Shayler, T. Lawa, and I. Pegg. The effects of early inlet valve closing and cylinder disablemention fuel economy and emissions of a direct injection diesel engine. In *Energy, Volume 79, Pages 100-110*, 2015.

Appendix

Appendix 1

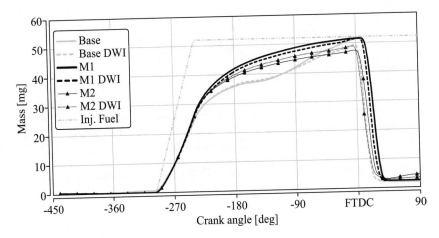

Figure A1.1: Fuel mass evaporation with different valve strategies

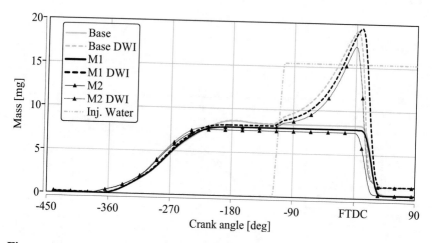

Figure A1.2: Water mass evaporation with different valve strategies.

Printed in the United States
By Bookmasters